How the Telegraph
Changed the World

ALSO BY WILLIAM J. PHALEN
AND FROM MCFARLAND

The Consequences of Cotton in Antebellum America (2014)

American Evangelical Protestantism and European Immigrants, 1800–1924 (2011)

How the Telegraph Changed the World

WILLIAM J. PHALEN

McFarland & Company, Inc., Publishers
Jefferson, North Carolina

LIBRARY OF CONGRESS CATALOGUING-IN-PUBLICATION DATA

Phalen, William J., 1942–
 How the telegraph changed the world / William J. Phalen.
  p. cm.
 Includes bibliographical references and index.

 ISBN 978-0-7864-9445-3 (softcover : acid free paper) ♾
 ISBN 978-1-4766-1867-8 (ebook)

 1. Telegraph—Social aspects—History—19th century.
 2. Telegraph—Economic aspects—History—19th century.
 3. Telegraph—United States—History—19th century. I. Title.

 HE7631.P46 2015
 384.109′034—dc23   2014043014

BRITISH LIBRARY CATALOGUING DATA ARE AVAILABLE

© 2015 William J. Phalen. All rights reserved

No part of this book may be reproduced or transmitted in any form or by any means, electronic or mechanical, including photocopying or recording, or by any information storage and retrieval system, without permission in writing from the publisher.

On the cover: worker repairing telegraph line, ca.1862. Photograph by Andrew J. Russell (Library of Congress)

Printed in the United States of America

McFarland & Company, Inc., Publishers
 Box 611, Jefferson, North Carolina 28640
  www.mcfarlandpub.com

For Maureen

Table of Contents

Preface 1

Introduction 5

1. Samuel Morse's Invention Through the Eyes of the Newspapers 13
2. Diplomacy 29
3. The Telegraph and the American Civil War 50
4. The Telegraph and Abraham Lincoln 74
5. The Atlantic Cable 89
6. The Telegraph and the Railroads 110
7. The Telegraph and Business 120
8. The Telegraph and the Press 128
9. Western Union 149
10. The Telegraph Workers and Unionism 171

Conclusion 185

Chapter Notes 189

Bibliography 203

Index 211

Preface

The earliest system of telegraphy for signaling over long distances is said to have originated in Africa. The means used were telephonic, and the signals were read by sound, and not by the eye, as is the case of the semaphore or other early signaling devices. The "Elliembic," as the instrument was called, produced a sound when it was struck that could be varied to produce a language.

Up until the nineteenth century, mankind communicated basically on a face to face basis. The attempts to send messages over long distances were usually limited by the elements (rain and wind would put out signal fires), by lack of daylight (semaphore messages could not be seen at night), and by line of sight problems (the illumination from a lighthouse is stopped by the curvature of the earth). Attempts to increase the ability to communicate over distance were limited by inventions and knowledge that had to be established first, in this case to understand electricity and how it could be used.

Most of the experiments using electricity were done in Europe, until the advent of Benjamin Franklin. His kite experiment which established his well-known name led to more experiments both in America and Europe using electricity. One was made by Franklin himself when he transmitted an electric spark across the Schuylkill River near Philadelphia.

In 1827, an American, Harrison Gray, erected a line two miles long around a racetrack, and almost succeed in establishing a true telegraph. What was missing was a battery; with that he could have anticipated Samuel Morse.

Another American, Joseph Henry, the secretary of the Smithsonian Institute, almost succeed in establishing a telegraph in 1828. He had devised and used a battery, but it was not strong enough to make his apparatus function.

Preface

While the invention of the telegraph waited on an understanding of electricity, those who almost made it work waited on first a battery and then one that would power the machine. The answer was the Daniell cell, a type of electrochemical cell invented in 1836 by John Frederic Daniell, a British chemist and meteorologist. His cell battery was the last piece of the telegraphic puzzle.

And then we come to the genius of Samuel Morse and this was in the discovery of the relay. "It was this discovery of a means by which the current, which, through distance had become feeble, could be reinforced or renewed by its own action. It made transmission from one point on a main line through indefinitely great distances, and through an indefinite number of branch lines, and to an indefinite number of stations, and registration at them all, by manipulation of a single operator at a single station, both possible and practicable."[1]

There were a number of men who contributed to Morse's success in the invention of the telegraph by doing work and developing ideas upon which he constructed his product, and an additional group that attempted to claim the credit that was justly his.

In 1839, Morse's attorney, Amos Kendall, wrote the following:

> Truth is slowly making its way against error. Even the adversary counsel in a late argument before the Supreme Court of the United States admitted that Professor Morse was the first to invent *a practically useful electro-magnetic marking telegraph*. The world will not hesitate to believe that which interested counsel do not think it expedient to deny.[2]

In this book, I will show the effect of an invention such as this on a new country—how, once the telegraph was perfected, its applications were seized upon by American entrepreneurs leading to new companies that spread this means of communication not only nationwide, but also worldwide beginning with a telegraph line under the Atlantic Ocean. The telegraph had a profound effect on business, allowing companies to operate more efficiently in their selling, marketing, and operations departments and ultimately allowing for the creation of one of the first of America's giant corporations—Western Union. The newspapers became more effective in relaying topical information because of the telegraph wire, and its use during the Civil War showed its effectiveness in battle.

Preface

    The wonders of the telegraph also had an effect on the American citizenry, who now did not have to only rely on the mails to gain and send information, and additionally received this information as it was sent. A battery, a telegraph key, and a piece of wire guided by Samuel Morse produced a revolution.

Introduction

In the town of Cobh in Country Cork, Ireland, sits the Cathedral of St. Colemans. The cathedral was overbuilt for the size of the town principally because of its location—on a hill overlooking the harbor of the City of Cork. During the famine of the nineteenth century as many Irish left Ireland by ship, they believed St. Colemans was the last piece of their native land they would ever have contact with. It must have been with sheer amazement that twenty-five years later, an instrument would be discovered that would reconnect them with their country of origin when the Atlantic Cable was successfully laid, allowing instant communication between their homeland and their new life.

This display of wizardry had been accomplished by the efforts of many men in many lands, beginning with simple communication by sight and sound and progressing to a method of allowing this communication at a distance, day or night, in any weather, through any obstacle.

In October of 1832, an artist by the name of Samuel Morse, travelling on the packet ship *Sully* from Harve to New York, entered into a conversation with, among others, Dr. Charles T. Jackson of Boston, Massachusetts. The topic of the conversation was electro-magnetism, which was the subject of a number of lectures the doctor had attended in France. After the doctor had described an electro-magnet to the group, he was asked "if the velocity of the electricity was retarded by the length of the wire." In reply, Dr. Jackson stated that electricity passed instantaneously over any known length of wire. Samuel Morse then remarked to the group, "If the presence of electricity can be made visible in any part of the circuit, I see no reason why intelligence may not be transmitted instantaneously by electricity."

Using this thought as a starting point, Morse began to construct what would become the telegraph. Several factors were already known

Introduction

to him because of the work that had been done by others. The instantaneous passage of the current or spark was one. The appearance of a spark when a current was interrupted was another. Since these factors were a given, he reached his first practical conclusions—that the current could be used to denote a sign; its absence, a second sign; and the time of the absence of a current, a third sign. From this point, the challenge was to construct signs which could be used to indicate letters. Still on the *Sully*, Morse the artist worked out his ideas on paper. This record, attested to by the ship's captain and passengers, would prove that what became the Morse alphabet (or code), simple arrangement of dots and spaces, was the invention of Samuel Morse.

Creating his telegraph was, however, only part of his eventual success. Now Morse had to prove that it functioned and patent his invention, both in the United States and abroad because there now appeared a plethora of rival inventors and inventions. Convincing Congress to give him a grant of $30,000, in February of 1843, Morse completed a telegraph line from Washington, D.C., to Baltimore, which proved that his apparatus was a success. At this point in the history of the telegraph as it pertains to Samuel Morse, there were two developments. First, Morse decided that the telegraph belonged to the nation, and so offered it to the federal government for the sum of $100,000. The government's decision to refuse the offer was to be part of many debates into the future as to who should control this means of communication. Also, when the offer was turned down by the government, Morse then went to the private sector to develop his invention.

The other development concerned Morse himself and his claim to be the inventor of the telegraph. In the Supreme Court Decision *Smith v. Downing*, Justice Woodbury held:

> From that time forward (1832), Morse is entitled to the high credit of making attempts to construct a practical machine for practical, popular and commercial use, which would communicate to a distance by electro-magnetism, and record cheaply and quickly, what was communicated, however imperfectly informed he may then have been of what had already been accomplished toward it, and he has the still higher credit among the experimenters of that time to 1837 of having then succeeded in perfecting what he describes at that time in his caveat and specification. Among about sixty-two competitors to the discovery of the electric telegraph by 1838, Morse alone, in 1837, seems to have reached the most perfect result desirable for public and

Introduction

practical use. By a lever deflected by a magnet, provided with a pen to write, with machinery to keep paper moving, so as to inscribe dots and lines, and more especially with an alphabet, he accomplished the great desideratum. Thus the fortunate idea was, at last, formed and announced, which enabled the dead machine to move and to speak intelligibly, at any distance, with lightning speed.[1]

During his lifetime, Morse would have occasion to fight off a number of claimants to his invention; this case was the basis of his right to his property. With the invention's worth proven, its uses began to multiply. First was the introduction of the telegraph to the railroads. When the railroads began to look at the telegraph as an adjunct to their operations, they immediately saw the value in the Morse system against its competitors. What was needed was a cost effective, simple means of communication, which could be utilized by all of the railroad's employees, and the Morse telegraph fit all the railroad's needs. More than that, both the railroads and the telegraph companies benefited. Once the system was in operation, the railroads gained a share of telegraphic revenues, coupled with a means of communication not only with their own widespread offices, but with other companies that they did business with, besides using the telegraph to coordinate traffic on the rails. Furthermore, the telegraph allowed the railroad to economize on their use of capital by avoiding double-tracking. With the telegraph lines adjacent to the tracks, the railroads provided greater market penetration, which increased telegraph demand. The telegraph companies thus acquired protection, economy, permanence, and strength. This combination of the railroad and telegraph is an example of two parties whose interests were mutual, cordial, and who provided for each other's necessities. Not only did rail and the telegraph act jointly to increase trade, they also contributed independently. Better rail connections meant cheaper freight rates regardless of telegraphic density. A denser telegraphic network would increase trade through better use of the rail network, but it could contribute to expanded trade through other channels as well.

Early in its existence the telegraph proved its worth in war, first in a small way in the Crimean War, then extensively in the American Civil War. It was part of the triumvirate of inventions—with the railroad and the steamship—that epitomized the material strength of the Federal war effort.

Introduction

The telegraph's obvious use in war is of course communication between the various parts of a nation's military in order to maximize its efforts against an enemy by coordinating its forces. The Civil War, however, brought out another use for this invention, that of the connection between the military and the populace, especially when this connection was broken. It was the constant telegraphic communication between the capital and the army that reassured the people, bound the loyal states together, and stimulated the civil and military authorities to greater exertions in contributing both manpower and funding to the Union cause. The telegraph for the first time gave the nation's commander-in-chief the ability to know what actions the military forces were engaged in, and the ability to change the outcomes of battles and the war itself. Ultimately, it was the chief medium between the government, the army and the people.

The impact of the telegraph on diplomacy can be seen in its effects on diplomats. At one time, a diplomat was an individual of great power, a true plenipotentiary. With the advent of a device that allowed a nation to communicate directly with another country without an intermediary, the power and strength of the diplomat began to wane. Also, since a nation's diplomatic efforts did not go through an intermediary, the speed and accuracy of the messages increased, as did diplomatic phrasing, because the language of a telegram was sharp, concise, and comprehensive.

The effect of the telegraph on the nation's newspapers was enormous. The first reason was speed. Before the telegraph, a newspaper usually had a combination of local news which was amassed by reporters on a daily basis, and foreign and national news that could be weeks old. With the telegraph, the age of the national news could now match the age of the local news. Because of this, readers fostered the use of the telegraph so that they could have the most up-to-date information available, which of course pushed the telegraph companies to increase their coverage and spurred the creation of new companies. This need for news operated in spite of warfare. Before 1861, the fastest method of sending a message from Missouri to California took ten days. After Western Union completed a line over the same route on October 24, 1861, the distance was covered instantaneously.

One thing lacking in the nation's papers was the inclusion of fresh

Introduction

foreign news. To incorporate the freshest possible world news, enterprising newspapers and organizations experimented with various ways to speed up the process. They would hire fast ships to meet the incoming vessels, use carrier pigeons to bring in the news before the ships would dock, and finally, build telegraph lines into eastern Canada to collect the foreign newspapers on board and then telegraph the information to New York, well before the ship would reach port.

With the telegraph as the news gatherer of the world, newspapers in New York City banded together to create an association of daily journals known as the Associated Press, which in conjunction with other groups of newspapers both domestic and foreign turned the city into the focal point of all news. Information would flow into the city by telegraph, where news of commerce, politics, war, and crime would be skillfully analyzed and re-distributed worldwide.

This system was initially incomplete until a man, who like Samuel Morse knew nothing about what he was about to undertake, completed the Atlantic cable connecting the United States with Europe. Cyrus Field was in fact the owner of a printing company in New York City who believed that he could reduce the time it took to send a message across the Atlantic from one month to a second or two. After a number of attempts, Field succeeded in connecting his cable, only to see it cease to operate, to worldwide dismay. Nine years later, in 1866, Field succeeded, establishing public confidence in cable communication.

With the Atlantic cable successfully laid, other underwater lines were constructed, bringing the world closer together, especially for trade. As an example, Sanford Fleming, a leading figure in transportation and communication in the late nineteenth century, made the following presentation to the Canadian Parliament in 1898 in support of a telegraph cable from Canada to Australia: "Steam and electricity go hand in hand. Trade and telegraphy are intimately blended, the latter being the most valuable ally of the former in building up and expanding all business. As a means of collecting cargoes for shipment, the telegraph is indispensable—without electric cables across the Atlantic the successful operation of fleets of ocean steamships would be impossible."[2]

An important link between the diffusion of the telegraph and the growth of world trade was its use to increase the efficiency of shipping by allowing ship owners to better allocate their shipping capacity. A ship represents a fixed cost. The more revenue generated per period, the higher the return. The telegraph allowed ship owners to coordinate supply and demand of cargo to update routes in response to changes during the ship's voyage, thereby increasing the value of each leg of the journey. Also, the telegraph allowed ships to potentially reduce time spent in port.

The impact of the telegraph on the transformation of business practices in the period when the United States was moving from an agricultural to a manufacturing economy was powerful. While I have already covered the relationship between the railroad and the telegraph, in fact the impact of the telegraph preceded the railroad in allowing business to expand. The transcontinental railroad was completed in 1861. Telegraph systems had been established in both the East and the West and reached Chicago, St. Louis, and New Orleans, as well as all the other principal cities of the nation by 1851. In this expansion, business demand for telegraphic services was pivotal. Although the telegraph came to be used locally until it was replaced by the telephone, in the public's mind the impact of the telegraph was on the shrinking of distance, rather than as an instrument for local use. Because of the telegraph's ability to increase a firm's business, a large number of local telegraph companies sprang up beginning in 1846 to service the growing number of small merchants, wholesalers, and speculators who wished to take advantage of the new technology.

As opposed to facing the usual difficulties that a new businesses encounters and which create a great many failed enterprises, the telegraph companies prospered. Companies small and large, including the banks, which came to rely on the telegraph, provided much of the necessary funds for the further development of the industry.

The smaller telegraph companies, however, had competition problems, since the lines of new companies were built adjacent to those of existing telegraph companies, in an effort to capture the most business possible. This could be done because the cost of starting a telegraph company could be relatively low depending on the area serviced.[3] By 1850, competition became the main cost factor, as mergers and acquisitions grew in the industry. As the companies became larger, the smaller companies became victims of promotional schemes and lower pricing which began to drive them out of business.

Introduction

The chief beneficiary of these mergers and acquisitions was the Western Union Telegraph Company. After its reorganization in 1854, It became so voracious in buying other telegraph companies, it was said Western Union, "embarked upon a career of conquest which has seldom been equaled in corporate history."[4]

This effort on the part of Western Union to consolidate the nation's telegraphic services was so successful that it attracted a number of opinions as to how the nation should deal with situation. One of those involved was Gardiner Hubbard, who organized the Bell Telephone Company in 1877 and who considered the power of Western Union to be simply "evil." He had two proposals to deal with the situation. The first was that the company be nationalized; however, the problem was that Western Union's value by 1883 was $80,000,000, which Hubbard realized Congress would not appropriate. This left the second option—creating a telegraph system within the nation's postal system. A bill in support of this idea failed, leaving Western Union as one of the largest monopolies in the nineteenth century.

1

Samuel Morse's Invention Through the Eyes of the Newspapers

In early 1841, both English and French telegraphers were looking for business in America. At the same time, Samuel Morse was attempting to have the Congress grant him $30,000 so that he could construct an experimental telegraph line. A Frenchman, M. Gonon, had asked the House Committee on Commerce for only $5,000 to test his idea to connect Washington to New York using twenty-seven semaphore stations.

During the summer and early fall of 1842, Morse, in an effort to prove that his concept was better than any others, set up demonstrations in New York City, including successfully passing an electric current through thirty-three miles of wire. Journalists from the New York *Tribune* witnessed one trial, and in an article referring to Morse's effort, described it as "among the most wonderful and, prospectively, the most useful applications of science to the great purposes of life which the present age had seen."[1] More important, they asked Congress to quickly grant Morse the appropriation.

One of these trials that Morse put together was before the American Institute,[2] which awarded him a gold medal and exhibited his apparatus at its annual fair. In a review the New York *Herald* predicted that Morse's telegraph would prove to be "the great invention of the age."[3] The newspaper also hoped that Congress would give Morse the money for his test.

Morse's next exposure to press reports of his invention came in October of 1842, when he and Samuel Colt, the inventor of the revolver, made a joint demonstration. Colt wished to demonstrate that gunpowder could be ignited by sending an electrical current under water. Morse became involved because in order for the telegraph to be useful, it would have to operate across American rivers. The experiment worked. An electrical charge was sent on a line under water to a mine on a ship in New York Harbor, blowing it to pieces. As the New York *Herald* described it. "Bang! Bang! Bang! Combusti-blowup eruption! ... 1,756,901 pieces."[4]

The same day Morse gave another demonstration of his telegraph to the American Institute, this time sending a message under water from Governors Island to the tip of Manhattan. The demonstration was inconclusive since his original battery was too weak and he had to exchange it for one of Colt's. Said the *Herald*, "neither entire failure, nor entire success."[5]

The next day Morse tried again with the *Herald* as his major booster. "All such may now have an opportunity of fairly testing it. *It is destined to work a complete revolution in the mode of transmitting intelligence throughout the civilized world*."[6] This time Morse had a short lived success—the transmission went through until a ship leaving port pulled up his cable along with their anchor.

In early December, Morse put together a demonstration in Washington, D.C. In this test, he ran a wire between the rooms of the House Committee on Commerce and the Senate Committee on Naval Affairs. With this telegraph in place, anyone who wished could send and receive messages. The results were extremely positive, at least according to the Washington press. The *National Intelligencer* commented, "This invention has truly been placed among the greatest of this or any age. The mind is scarcely prepared to pursue even in speculation the mighty results which are soon to follow its practical demonstration."[7]

On February 23, 1843, Morse's appropriation bill passed the House, mainly because the Whig majority favored internal improvements, and on March 3 it was unanimously passed by the Senate and signed by President John Tyler. Morse began work immediately, deciding to run a telegraph line from Baltimore to Washington alongside the route of the Baltimore and Ohio Railroad, hoping to finish in time for the opening of the next session of Congress in December.

When difficulties prevented Morse from finishing the telegraph line on schedule, he placed a notice in the *Journal of Commerce* explaining the delay: "In an enterprise so entirely new, it can hardly be expected that every part can be conducted with the precision and perfectness

1. Morse's Invention Through the Newspapers

which is gained only by experience. Unforeseen difficulties will be encountered and are to be overcome, and delays will of course be incurred. There are no intrinsic ones as yet of a nature to shake the confidence of the most sanguine in the final triumph of the enterprise."[8]

On May 24, 1844, Morse's telegraph line was finally completed and the now famous transmission "What hath God wrought!" was sent. Morse followed this up with an article in the New York *Observer* headed "The Electric Telegraph Triumphant," explaining what had transpired.[9]

By May 27, The Democratic Party held its presidential nominating convention in Baltimore. Since there was intense interest in the proceedings, Morse took advantage of it and arranged to have coverage of the convention telegraphed to Washington, which gained enormous publicity for the telegraph. A Washington correspondent of the New York *Herald* reported that "little else was done here but watch Professor Morse's bulletin from Baltimore, to learn the progress of doings at the convention."[10]

The next day Morse created more excitement by making his telegraph part of the process for selecting candidate James K. Polk's running mate. Telegrams flew back and forth between the convention and Washington to the convention's possible contenders. The long-distance political bargaining, the *National Reporter* said, went on *"with lightning speed."*[11]

To further the publicity and to create an acceptance of his invention, Morse began to tell the public about how the telegraph worked and how it would be useful to Americans—who were fascinated by the telegraph—even to the extent of explaining how his conductors, galvanic batteries, and even his dot-dash code operated. "Is this mysterious power a substance or an effect?" asked the New York *Daily Times*.[12]

Bafflement was the common reaction to Morse's telegraph. It seemed to operate by "an almost supernatural agency," one newspaper said; "We stand wonder-stricken and confused." Americans compared it to a bottle-imp, a spell, a classical myth, something from *Arabian Nights*. The wonderment stemmed in part from the awesome harnessing of power they believed the telegraph represented.[13]

In forecasting social changes that Morse's telegraph might bring, antebellum prognosticators adapted their thinking to the Gospel of Progress. The New York *Sun* proclaimed the telegraph "the greatest revolution of modern times and indeed of all time, for the amelioration of Society." It would create civic order, strengthen domestic ties, bring harmony among nations, and redeem mankind.[14]

In this spirit, the Utica *Gazette* anticipated an "immense diminution" in crime. Felons would give up hope of escaping justice: "Fly, you tyrants, assassins and thieves, you haters of light, law, and liberty, for the telegraph is at your heels."[15] Domestic joy and sorrow would thrill along the wires: "The absent will scarcely be away," stated the Philadelphia *North American*. "The mother may, each day, renew her blessing upon her child a thousand leagues away; and the father, each hour, learn the health of those around his distant fireside."[16] Ultimately nations would be wired to each other, making the planet a neural map, what the *Christian Observer* called a "sensorium of communicated intelligence."[17] As the New York *Herald* put it, "What a future!"[18]

This new era of communication begun by Morse and his telegraph would overcome those who resisted its blessings. According to the New York *Herald*: "Steam and electricity, with the natural impulses of a free people, have made, and are making, this country the greatest, the most original, the most wonderful the sun has ever shone upon.... Those who do not mix with this movement—those who do not become part of this movement—those who do not go on with this movement—will be crushed into more impalpable powder than ever was attributed to the car of Juggernaut. Down on your knees and pray."[19]

And to Morse, all glory: "This eminent individual—the inventor of the last and greatest wonder of the age—is under forty-five. He is a little above the common height, and rather thin; his hair slightly grey, complexion dark and sallow, eyes brilliantly black, with a peculiarly soft and gentle expression."[20]

In an effort to find out where his genius came from, the *Home Journal* subjected Morse's head to a phrenological analysis. The newspaper told its readers that Morse's well-developed organs of Constructiveness and Identity revealed him to be hard and soft: forcible, persevering, almost headstrong, self-relying, independent, aspiring, goodhearted, and eminently social, through sufficiently selfish to look well to his own interests."[21]

After the success of the Baltimore-Washington telegraph line, Morse began to think about his next move with the telegraph. He

1. Morse's Invention Through the Newspapers

decided the best course of action was to sell it to the federal government. The press also looked into his future and the question of whether or not the telegraph should be nationalized. The Washington *Union*[22] argued that the government would distribute the benefits of the telegraph impartially, because in private hands it could become a "dangerous monopoly." Newspapers such as the New York *Evening Mirror*, however, believed the opposite; it accused the *Union* of regarding the people as "a set of knaves and swindlers, and the officials of the government as being alone worthy of holding any trust." The *Mirror* believed that congressmen would misuse the telegraph for their private convenience at public expense: "The Hon. Mr. Hopkins will send word to his wife in Buffalo that he had a comfortable night's sleep."[23]

Morse then petitioned Congress for more money to extend the Washington-Baltimore telegraph line to New York City. While waiting for a decision, Morse continued his experiments, one of which was running telegraph lines under water. Such was Morse's reputation with the press that the *Journal of Commerce* reported, "It is even hoped that a telegraphic communication may be made with Europe, and at no very great expense."[24]

When the Congress finally decided the fate of the new telegraph line to New York, the news was not good. The line extension was defeated, and in its place was a salary for Morse for the upkeep of the Baltimore-Washington line. With no public money available, Morse had the good fortune to meet Amos Kendall, postmaster general under President Andrew Johnson and a man who understood business, which Morse did not. Kendall immediately began to form telegraph companies, one of which was in partnership with Henry O'Reilly, an Irish immigrant who was also interested in expanding the telegraph.

With his business interests under control and a cash flow derived from the new companies, Morse looked to expanding his invention in Europe. The problem here was that London journals such as the *Globe* and the *Literary Gazette* wrote that his telegraph was not needed, that everything that he had accomplished in his Washington-Baltimore line had been done earlier and better by an English inventor, Charles Wheatstone. "The English reader need scarcely be informed, "said the London *Mechanics Magazine*, "that Mr. Morse has ... only *re* discovered what was previously well known in this country."[25] American newspapers in return printed articles such as "The Magnetic Telegraph—*American and British*" giving the credit to Morse: "The Electro-Magnetic Telegraph of Professor Morse is the first realization of a practicable Telegraph on the Electric principle," one wrote; "Thefts of new inventions and bold plagiarisms of whole works, the offspring of American minds, have been so common in England, that we deem it almost a matter of course that no acknowledgement will ever be made in English works of indebtedness for anything of the kind to America."[26]

Morse looked into Wheatstone's apparatus and found it to be overly complicated. As the London *Pictorial Times* indicated in an article, it became difficult to use: "The left hand needle moving to the left twice gives *a*, three times *b*, once to the right and once to the left *c*.... The order is then taken up by the right hand needle, moving once to the left for *h*, twice for *l*, The remaining signs are made by the two needles working conjointly, so that the simultaneous movement of the two, once to the left, indicates *r*, twice for *s*, three times for *t*, once to the right and once to the left for *u*, once to the right for *w*, twice for *x*, and three times for *y*."[27]

Even though his telegraph was superior to Wheatstone's, Morse was unable to make any headway in England. He therefore travelled to Paris where he encountered a Wheatstone telegraph modified by a French scientist, Louis Breguet. Morse attempted to convince the French Chamber of Deputies that his system was better, but again encountered a nationalistic newspaper, the *Revue Scientifique*, which said about his telegraph: The way the signs are produced, the writing carried out at a distance, the ... composing of the alphabet, all these elements are essentially defective, and it will be necessary to abandon them."[28]

Returning to the United States, Morse began work on the New York to Washington line. Before spring in 1846, the New York *Observer* announced that it would be possible to speak by telegraph directly from Boston to Washington.[29]

Before the line could be completed, on May 12, Morse's fortunes received a huge boost when the New York *Sun* printed an exclusive: the entire text of President Polk's address to Congress of the day before requesting a formal declaration of war against Mexico and the authority to call up troops. The speech took more than two and one half hours to transmit and was by far the lengthiest document ever sent by telegraph.

1. Morse's Invention Through the Newspapers

Morse immediately sent a copy of the *Sun* to his friend Dominique Francois Jean Arago, the permanent secretary of the elite national scientific society, the Academie des Sciences, which presented the newspaper to the Chamber of Deputies. Impressed, the chamber voted an appropriation of nearly half a million francs.

A month later, the line to New York was almost complete. The Manhattan line stopped in New Jersey because the problem of running an underwater wire under the Hudson River was still not solved. However the Washington *Union* reported part of a four way conversation between Morse, Alfred Vail (one of his partners), and two operators from four different points—New Jersey (New York), Philadelphia, Washington, and Baltimore, over 260 miles:

> Washington.—Baltimore, are you in contact with Philadelphia?
> Baltimore.—Ay, ay, sir; wait a minute. (After a pause.) Go ahead. You can now talk with Philadelphia.
> Wash.—How do you do, Philadelphia?
> Phila.—Pretty well. Is that you, Washington?
> Wash.—Ay, ay; are you connected with New York?
> Phila.—Yes.
> Wash.—Put me in connection with New York.
> Phila.—Ay, ay; wait a minute. (After a pause.) Go ahead. Now for it.
> Wash.—New York, how are you?...
> New York.—Ay, ay Washington, write dots. (Washington begins to write dots.) That's it; O.K. Now I have got you.[30]

With these lines completed, the telegraph began to spread out especially in a western direction. O'Reilly not only expanded in this direction, but also looked for improvements in the system. One was an invention by a Vermont inventor, Royal House. His machine used a key board, stamping one key for every letter. The New York *Herald* said of it, "Instead of an arbitrary character, like that which is used in Professor Morse's machine, there is a letter of the alphabet. It is decidedly superior to any other telegraph ever used."[31]

After Morse examined House's machine, he dismissed it as being too complicated. An editor of the New York *Evening Post*, however, told Morse that the device was important.

Adding to his problems, O'Reilly attacked Morse as a monopolist and was so successful that the press finally turned against him. Newspapers had now become dependent on the telegraph for their news stories and were looking for lower rates from O'Reilly than from him. New York papers stood to gain the most from a rate war between them, especially Horace Greeley's New York *Tribune*, because of the amount of telegraphic information that the paper consumed. As a result, the *Tribune* led the attack on Morse, "Shall [the telegraph] be republican and free or an agent of aristocratic despotism—shall it be American or shall it be Russian?"[32] One of the papers that defended Morse was the *Day Book*, which accused O'Reilly of bribing editors with gifts of stock. The paper stated: "He has made half the editors in the country his partners. Why shouldn't they think of him one of the greatest men of the age?"[33]

The battle between O'Reilly and Morse intensified when O'Reilly began to challenge Morse's patents, and whether or not he should have received the money from the government to build the Baltimore-Washington line. A Louisville newspaper supported O'Reilly by attacking Morse: "Here is a deliberate falsehood, a palpable and blushing fraud practiced on the Representatives of the people for the purpose of getting thirty thousand dollars voted to Professor Morse and his associates."[34]

The subject of the conflict then became one of "general principle," in other words, according to O'Reilly, the government had granted Morse ownership of a force of nature, in this case electromagnetism, which could not be patented. The *Daily Post* in New York supported O'Reilly's argument: "Suppose a man who had obtained a patent for a water wheel to propel machinery, were to set up the claim that no one could use water to propel machinery by any kind of wheel subsequently invented, though it might differ from his in every particular, would not even Mr. Morse treat with contempt so absurd and preposterous a claim; and yet, it is no more preposterous than his attempts to monopolize Electro Magnetism."[35]

Morse's replies to O'Reilly's attacks were generally low key, because he believed that O'Reilly's course of action was self-destructive. In a letter to the editor of the Louisville *Journal*, Morse wrote, "Through these reckless attacks upon me, the merits of inventions, rightfully belonging to our own country (as in the present case) is given to other nations."[36]

Finally in the spring of 1848, Morse had had enough and sued O'Reilly over patent infringement. The deliberations went through the

1. Morse's Invention Through the Newspapers

summer, and on September 13, Morse won the case so convincingly that The Frankfort *Yeoman* commented, "So broad and comprehensive that nothing is left for evasion, O'Reilly will see he has no means of going on."[37] O'Reilly kept fighting, however, supported by newspapers such as the Louisville *Courier* that attempted to inflame public opinion against Morse. "From one end of this Union to the other the people will speak out on this subject, and this monster can never now be snuggled through Congress, it is dead, dead, dead."[38]

O'Reilly tried again to best Morse by purchasing the new Bain telegraph, which used a system of recording the message beforehand on a disc. Once the disc was perforated, the entire message was sent at a speed much faster than Morse's telegraph could send it. The judge's decision in *Bain v. Morse* was ambiguous, with both sides claiming victory. The press basically supported Morse. A Rochester paper criticized Bain's machine as a "cast off British affair"[39] and the New York *Sun* agreed: "We go in strong for our own country. We are determined ... that American inventors shall not be deprived of their just fame."[40]

These court cases that basically challenged Morse's position as the inventor of the telegraph continued until what became known as "The Great Telegraph Case" came before the United States Supreme Court in December of 1852. The case was brought before the court by O'Reilly in support of his telegraph. The two questions before the court were: was Morse the first and original inventor of the telegraph, and was O'Reilly's telegraph substantially different from it.

The following editorial in the New York *Times* summed up Morse's position:

> Grant that MORSE was, as is claimed, indebted to the suggestions of others.... MORSE was *the man* who was publicly experimenting, in our midst, on this subject—inviting scientific gentlemen to witness his progress, who besieged the doors of Congress for an appropriation to enable him to demonstrate the practicability of his invention; who entered his caveat and obtained his patent; who, in 1844, laid down the first line of Electric telegraph in this country, from Washington to Baltimore, and sped the first aerial message on its electric path.... If others knew that electricity could be used for recording language at a distance, and kept the knowledge from the public, or were too indolent or careless to reduce it into practice, we think they are too late to claim the credit, after another, by labor and devotion, has accomplished the work.[41]

How the Telegraph Changed the World

The decision gave Morse the credit for the invention of the telegraph. However, on Morse's claim to be the owner of electro-magnetism, the court found this to be too broad because some inventor could develop a better way of telegraphically recording by electricity without using Morse's system. Morse considered the opinion a validation of his patent. Additionally, after he applied for a seven year extension of his original patent, it was granted by the Patent Office.

In 1854, Morse received a letter from a New Yorker named Cyrus Field concerning his idea of running a telegraph cable across the Atlantic. With his court cases behind him and the possibility of becoming involved in a new venture before him, Morse not only agreed with the project, but joined financially. As with any new public adventure, the newspapers had an opinion. One America paper stated in an editorial: "All idea of connecting Europe with America by lines extending directly across the Atlantic is utterly impracticable and absurd."[42] Said the New York *Tribune*, "It is impossible to contemplate the probability of such an achievement without a glow and a thrill at its sublime audacity and its magnificent uses."[43] And the New York *Observer* called it "a more sympathetic connection of the nations of the world than has existed in history."[44]

While waiting for the laying of the Atlantic cable to begin, Morse became involved in politics. He ran for Congress in 1854, and although losing, his showing was respectable. Morse ran as a Democrat, but was also a member of the Council of the Order of United Americans—the Know-Nothings, whose ideas he supported.

This support led Morse to begin an eight month fight in the press with the Catholic bishop of Louisville, M. J. Spaulding. At issue was a statement allegedly made by the Marquis de Lafayette: "American liberty can be destroyed only by the Popish clergy." This warning had been quoted since it first appeared in Morse's anti–Catholic book, *Confessions of a French Catholic Priest* (1837).[45]

Just before Morse's 1854 run for election, an article reprinted from the Cincinnati *Enquirer* appeared in many newspapers claiming that Lafayette's statement was taken out of context. According to the *Enquirer*, what Lafayette actually said was, "the fears … that if ever liberty of the United States is destroyed it will be by Romish priests—are certainly without any shadow of foundation whatever."[46] Morse soon became dis-

1. Morse's Invention Through the Newspapers

illusioned with the Know-Nothings just after the news of his battle with Bishop Spaulding crossed the Atlantic. The Edinburgh *News of the Churches* reported: "By the world-renowned Professor Morse, Bishop Spaulding is pilloried in Kentucky."[47]

Finally, the cable was to be laid to complete the Atlantic side of Field's two prong attempt to connect Europe with America. With Morse as a witness, a gale struck, causing the cable to be lost overboard. The second problem concerned Morse's relationship with Field. After returning from the cable laying disaster, Morse learned that Field had purchased a rival telegraph made by David Hughes. Field explained that he had done it so that it could not be purchased by someone else and used in competition with the Morse system.

While Morse wished to stay in good graces with Field, he learned that the New York *Tribune* had published a piece entitled "Astounding Telegraphic Improvements," and spoke of Hughes' telegraph as "the most wonderful instrument for telegraphic purposes ever invented ... far ahead of any machine now in use." The article also reported that Field had begun a new company, the American Telegraph Company, to build Hughes lines, "a network of wires radiating in all directions from New York to every prominent business place in the Union."[48]

Morse travelled to the Continent as preparations for the laying of the Atlantic cable were underway, visiting France, Denmark, Russia, and England to find that his reputation and his telegraph were held in high esteem. The London press praised Morse and the cable as being healers of old wounds. The London *Times* reported, "The guest at the Albion has deserved well of the world, and in his generation has done much to advance the cause of human progress. We rejoice to see England and America united in a project so honourable to both nations."[49]

After the Atlantic cable had been laid,[50] the celebrations began; the American people were greatly elated. The New York *Tribune* announced: "A mighty though silent transformation in the conditions of human existence had just been effected, we have been thrown into the immediate intellectual neighborhood of the whole civilized and a large portion of the semi-barbarous world."[51]

Morse also gained his share of adulation; the Christian evangelical press celebrated Morse as one of their own, especially the *Western Episcopalian*, which described More as a humble Christian and a man of

God who now held "the highest position ever attained by mortal man uninspired.... Kings and emperors sink before him."[52] In Paris, he received great applause for the Atlantic cable and a dinner at which he was continuously praised. The New York *Times* said of his reception, "Every figure of rhetoric was exhausted in his praise, no man ever received a greater ovation from his fellow human-beings."[53]

The onset of the Civil War conflicted Morse. Although his sympathies were with the South, he did not want war. To that end, he became involved in politics, attempting to bring both sides together. When Abraham Lincoln issued the Emancipation Proclamation on January 1, 1863, Morse believed it to be both unconstitutional and a blow to the South.

In February of that year, Morse attended a meeting of conservative Democrats at Delmonico's restaurant in New York. The purpose of the meeting was to revoke the Emancipation Proclamation. They would work towards that end by appealing over the president's head directly to the people, gathering northern support for the idea of preserving the Union without making Unionism a vehicle for abolition.[54]

The *Evening Post*, a newspaper that supported abolitionism, either sent or smuggled a reporter into the meeting who wrote several articles about what transpired. The articles depicted a "reactionist conspiracy," a "secret caucus," and a "brotherhood of Carbonari." There on the luxurious seats of parlor No. 4 were August Belmont, "A Hebrew from Germany," and Samuel F. B. Morse, "artist and inventor born at Charlestown, Massachusetts"—a traitor, the tag implied, to the patriotic heritage of New England.

The newspaper gave a sinister slant to the group's aims: "The rich men of New York are to supply the money ... for an active and unscrupulous campaign against the government of the nation, and in the behalf of a body of rebels now in arms." The purpose of the group was to raise money "to hand the government over, if they can, to the malignant slaveholding oligarchs who for nearly two years have been slaughtering our sons."[55]

Morse did not shy away from the group, but embraced their ideals by accepting the presidency of the organization whose formal name was the Society for the Diffusion of Political Knowledge. While the society presented its views on such subjects as civil liberties and states' rights, its main point was the defense of slavery. The society's main method of spreading its beliefs was through pamphlets, many of which Morse wrote himself.

1. Morse's Invention Through the Newspapers

Morse was gratified that his writings on the defense of slavery reached a wide audience—"some of the most pious, as well as distinguished, intellectual minds in the country." A southern paper, the West Virginia *Intelligence,* praised Morse as a communicator—first technological and now political: "There is a great fitness that the distinguished originator of the American Magnetic Telegraph, whose genius has chained the lightning and made it an obedient messenger to carry information with the rapidity of thought from end to end of the land, should be among the foremost to flash the light of political knowledge into the minds of his fellow countrymen."[56]

However, some papers did not agree with Morse; the New York *Times* mocked his views on human equality, not for holding that one body of men might be less able than another, but for deducing that in such a case "it is the right of the latter to rob, beat, and sell the former."[57] The *North American Review* derided as "worthless and shallow" his treatment of slavery as merely a form of government, his endorsement of the cornerstone doctrine, his sneers at the Declaration. The journal attributed these and his other notions to "the self-conceit of a sick man."[58] A Boston newspaper recommended that he be imprisoned.

As the war continued, the Western Union Telegraph Company, in its pursuit of becoming the nation's largest telegraph company, bought several smaller telegraph companies, many of which Morse had large investments in, making Morse even wealthier than he was before. According to the New York *Tribune.* These mergers were making telegraphy "the most profitable business in the country."[59] And as far as Morse was personally concerned, it wrote about his "regal income."[60]

When Lincoln's first term was coming to a close Morse backed McClellan, believing that McClellan would stop the war by making a deal with the South that included the continuation of slavery. At a meeting of the Democratic Party in Manhattan to which he escorted McClellan, Morse was introduced as "a gentleman who, by his scientific researches and discoveries, had made his fame and name immortal."[61]

After more than a quarter century of use, Morse's telegraph remained for many people an astonishment—the New York *Herald*[62] termed it "the greatest triumph of the human mind, the most direct proof of man's conquest of the human mind."

Morse travelled again to Paris and upon his return began the last years of his life unhappily. His home had been burglarized; his brother disappeared and was later found dead. Even a gala dinner in his honor organized by Western Union was a trial, even though the New York *Times* described it as "one of the most magnificent affairs of the kind that ever took place in this City."[63]

The dinner, however, had a purpose; a key speaker, William Orton, the president of Western Union, rose, not only to praise Morse, but also to attack the idea that the federal government should take over control of the telegraph. When Morse spoke, he recounted his early experiences in dealing with Washington. How he was given the money to conduct his experiment, but after its success, could not interest the government in purchasing it.

Morse's recollections were strenuously castigated by the New York *Herald,* which favored government control of the telegraph. In several editorials, the *Herald* questioned the intent of banquet: "It is difficult … to resist the conviction that the affair was got up less for the purpose of honoring Professor Morse than of advancing the interests of the Western Union Telegraph Company." If proof were needed, the *Herald* said, William Orton's "execrable"[64] remarks made it clear that the homage to Morse was propaganda for an offensive, "a decoy duck to affect a great lobby movement upon Congress." The revered guest and the influential diners had been bought off, "to spread abroad the erroneous impression that the professor and the company assembled to meet him were all opposed to the absorption of the telegraph in the postal system."

In 1871, Morse was again under attack by old enemies, especially O'Reilly and F.O. Smith, with whom he had dealings with early in his career. In February Archbishop Spaulding died. The New York *Herald* published a eulogistic letter to the editor recalling the prelate's newspaper duel with Morse eighteen years before. Morse had attributed to Lafayette the remark that "American liberty can be destroyed only by the Popish clergy." Spaulding had countered with a pamphlet and newspaper articles claiming that the partial quotation reversed the meaning of Lafayette's full statement, which was that Americans had no reason to fear Catholicism. His admirer in the *Herald* now recalled that Spaulding's power arguments, "the rude force with which his blows fell," had compelled Morse to retract. Morse saw the letter and in his greatly weakened state wrote a five-and-a-half-page retort—a full column of print presenting

Professor Morse sending a message, 1871 (Library of Congress).

a detailed case for attribution. "I retract nothing," he said. He mocked Spaulding's eulogist for thinking that attempts to defame him would go unnoticed. "He may have supposed I was dead."[65]

Morse passed away on April 2, 1872. The New York *Times*, possibly having some idea of the problems he faced at the end of his life, stated editorially that "the vexations and annoyances, the troubles and sorrows of the last few weeks of his life … contributed, in a great degree, to bring on his last fatal attack."[66]

Newspapers nation-wide hailed him as a great American. The Louisville *Courier-Journal* said, "If it is legitimate to measure a man by the magnitude of his achievements, the greatest man of the nineteenth century is dead."[67] *Patent Rights Gazette*: "The first inventor of his age and century is dead!"[68] New York *Herald*: "Morse was, perhaps the most illustrious American of his age."[69]

So great was the outpouring of affection for Morse that few papers, such as the New York *Golden Age*, mentioned his distrust of immigrants, hatred of Catholics and support of slavery, "Among the many … apologies for American slavery are some shameful passages from his pen."[70]

# 2

Diplomacy

Most Americans know of the Battle of New Orleans from their study of American history not only because it was a major victory over the British during the War of 1812, but also because the battle was unnecessary. The war had ended on December 24, 1814, after the British and the Americans signed what became known as the Treaty of Ghent in Belgium. The battle took place on January 8, 1815. The reason that for the continued fighting was of course the distance between Belgium and the United States and the time that it took to carry news between the two nations.

What is little known is the fact that with an increase in the speed of communications—this time between London and Washington—there is a great possibility that the War of 1812 would never have occurred. Of the list of the causes of the war, one of the most important was the "Orders in Council." The orders were a series of regulations that required Americans trading with Continental Europe to stop at Britain first, receive British permission to proceed, and pay a fee. Americans believed that the order was aimed at destroying its mercantile prosperity, and viewed it as Britain's attempted continuation of America's colonial status a generation after the Revolution.

While the Orders were a problem for the United States, they also had a deleterious effect on Great Britain because the United States Congress had passed a Non-intercourse Bill which produced a damaging boycott of British goods, leading to an economic depression in Britain.

On June 16, 1812, one day before the U.S. Senate passed the war bill, and two days before President James Madison signed it, Viscount Castlereagh, the British foreign secretary, introduced into Parliament a motion calling for the repeal of the Orders. He announced "that a proposition should be made to the American government to suspend immediately the Orders in Council, on condition that they would suspend their Non-intercourse Act." The revocation of the Orders brought rejoicing throughout Britain; it seemed that the American market would soon be open again to British goods. Instead, to their horror, British industrialists and workers learned six weeks after Castlereagh's announcement that the repeal had been too late; the United States had declared war. Ships carrying the British concession had crossed others carrying news of war.[1]

Many scholars have argued that a telegraph cable across the Atlantic would have allowed Americans to learn of Castlereagh's revocation of the Orders in time to prevent the War of 1812. In fact, twenty years later President Madison contended that if Britain had rescinded the Orders a few weeks earlier, "war would not have been at the time declared, nor is it probable that it would have followed, because there was every prospect that the affair of impressment + other grievances might have been reconciled after the repeal of the obnoxious orders in council."[2]

The lack of speed in long-distance communications[3] was not a new problem. The tempo of domestic politics had often exceeded that of international politics. Why were the consequences so disastrous? The answer has to do with the solutions that statesmen found to the difficulties of coordinating foreign policy across great distances in the age before electric communication. In general, foreign ministers compensated for slow communication by giving diplomats greater autonomy. This autonomy, once the telegraph was in use, began to diminish when foreign ministries began to oversee and direct the activities more easily. Diplomats had mixed feelings about the encroaching power of foreign ministries. Some bridled at efforts to centralize control over foreign policy. Many appreciated increased guidance. One example concerned the American minister to France, Richard Rush. On February 24, 1848, Rush reported to Washington via ship that revolutionaries had toppled King Louis Philippe's regime and erected a republic in its stead. Rush was immediately asked by the new government to recognize it in his capacity as the American minister. Rush was weeks away from his superiors He felt acutely isolated as he faced what he saw as a momentous decision. On the one hand, the U.S. government had specifically accredited Rush to the French court. He possessed no official status vis-à-vis the revolutionaries. On the other hand, the maintenance of good relations with the de facto government of France was the main goal of Rush's mission.

2. Diplomacy

On February 28, 1848, he unilaterally recognized the new regime. Justifying his action afterward to the secretary of state, he contended: "In recognizing the new state of things, as far as I could without your instructions, and in doing it promptly ... I had the deep conviction that I was stepping forth in aid of the great cause of order in France and beyond France—and that I was acting in the spirit of my government and country, the interpreter of whose voice it fell upon me suddenly to become. If I erred, I must hope that the motives which swayed me will be my shield. The provisional government needed all the moral support obtainable.... In such an exigency, hours, moments were important."[4]

Even as Rush pondered his response to the 1848 French Revolution, technological changes were bringing closer to realization the possibility of electric communication across the Atlantic Ocean. Among these changes, two deserve special attention: the steamship and gutta-percha. The first scheduled steamship lines crossed the Atlantic in 1838. Steamships, unlike vessels under sail, could reliably lay cable in a straight line at a controlled rate. Gutta-percha, a rubbery substance obtained from the gum of gutta trees, was used to insulate and protect the copper core of the cable. With these advancements, a cable was laid connecting the United States and Europe in 1866.

By 1870 there were three underwater telegraph cables spanning the Atlantic Ocean when Elihu B. Washburne, the U.S. minister to France, witnessed the next French revolution. On September 5, Washburne reported by telegraph from Paris that a republic had been proclaimed and asked his superiors for instructions. The next day, Acting Secretary of State J.C.B. Davis cabled that Washburne should "not hesitate to recognize" the provisional government as soon as it appeared to be the de facto source of authority.[5] Washburne officially recognized the Third Republic one day later on September 7, 1870.

At first glance the American responses to the 1848 and 1870 revolutions suggest that the Atlantic cable had only beneficial effects on the conduct of U.S. foreign policy. Washburne, like Rush, knew that he was implementing the wishes of his superiors. The State Department could respond much more quickly during the 1870 uprising than it had in 1848. Just as important, the Atlantic cable freed policymakers from the vagaries of transatlantic steamers. The shipping schedules caused American diplomacy to become a system composed of long waits punctuated by hurried responses so that the return message would make the next boat. After waiting weeks to hear about the outcome of the 1848 revolution in France, when Rush's message arrived, President Polk hurriedly called a cabinet meeting to discuss the events in order to send Rush instructions by the next steamer. At times, the president abruptly ended his letters with such phrases as "I cannot add more and save the mail."[6]

Obviously, the telegraph shifted control over foreign policy from the local minister to the policymakers in Washington. Rush's distance from the Secretary of State gave him autonomy and seemed almost to necessitate decisive action when faced with an immediate decision. In such a circumstance, a rapid communications capability changes who makes the decision. As an example, on January 14, 1848, Rush reported the death of the king's sister. The message reached Washington a month later. President Polk sent his condolences in a letter addressed to his "great and good friend," King Louis Philippe. By the time it reached Paris on February 29, the king had been overthrown. Rush decided to suppress the message from Polk since he was pursuing a policy of support for the new regime. By contrast, in 1870, the Prussian ambassador to France asked Wickham Hoffman, the secretary to the U.S. delegation in Paris, to assume protection of the North Germans in France during the Franco-Prussian War. Hoffman, then leading the legation in Washburne's absence, felt that this offer served the interests of the United States and believed that his government would respond favorably. However, he could not give the Prussian ambassador an answer. His reason, as he stated later, "If there were no cable across the Atlantic, and it was necessary to say 'Yes' or 'No' at once, I should say 'Yes'; but as there was telegraphic communication, and I could receive an answer in forty-eight hours. I must ask instructions from [the secretary of state]."[7]

As these examples show, the telegraph reduced the latitude that foreign ministers had over their actions and in some cases slowed the execution of U.S. policy by diminishing individual initiative. The reduced autonomy of American diplomats stationed in Paris deviates from generalizations made about the telegraph's influence on communications within late nineteenth century European colonial empires. The historian Daniel Headrick argues that the telegraph initially did not increase the control of the central government over imperial agents. While he was referring to a nation and its colonies, his point also applies

2. Diplomacy

to a situation in which a nation is in communication with one of its foreign ministers:

> The impact of the telegraph on colonial power relations is ambiguous, however. As the expansion of telegraphy coincided with the new imperialism, uncertainty on the frontiers of empire, ambitious imperial agents, a lack of information, and an overwhelmed central administration combined to enhance the power of the men on the spot at the expense of their home governments. But this was a temporary phenomenon associated with the phase of warfare and expansion. As soon as areas were pacified (or both the home government and its ministers became used to the telegraph), bureaucratic controls replaced the free-wheeling agents of the frontier period. And inevitably the controls operated through the telegraph wires and cables. After the turn of the century, one no longer hears much about independent proconsuls.[8]

There was also the question of having a variety of sources of information. In the instance of a colonial situation, the local representative could monopolize their end of what was generally a single telegraph line, enabling them to supervise the outgoing information that otherwise might have allowed their superiors to question and challenge their judgment. In contrast, if the situation concerned the information flow between a country and its representative to another developed nation, the foreign minister had no hope of gaining similar control over the communication because of the activities of journalists, businesspeople, and non–American diplomats with access to the same cables.

By the late nineteenth century, as an example, too many channels of communication connected the United States and Europe for a diplomat to gain a free hand because of the ignorance of his superiors about the situation involving the diplomat. The only answer that a diplomat had was to be the only "official" voice in a situation. In this circumstance, the chief foreign minister could monitor and control the sending or reception of messages at his end of the line. More than that, the cost, secrecy, and aura of importance surrounding telegrams made foreign policy officials anxious to prevent those outside the embassy from gaining information, and to limit the use of the telegraph to high-ranking bureaucrats. In 1875 the U.S. secretary of state decreed: "No messages are to be sent by the telegraph except those from the Secretary, or one of the Assistant Secretaries, or the chief Clerk, or those authorized to be sent by one of the above. Any Chief of Bureau or Clerk wishing to transmit any message on public business, by the telegraph can do so by obtaining the endorsement of the chief clerk." Likewise, the British Foreign Office restricted second-division clerks from processing telegrams until 1911.

Pressures towards greater hierarchy also affected overseas legations. In 1909 the U.S. Department of State decreed that only the head of mission and first secretary were allowed to handle the main codebook used for decrypting secret telegrams. In France the foreign minister asked heads of mission to provide written justification for each telegram they sent.[9] While in practice such regulations often proved unworkable, John Kenneth Galbraith, as U.S. ambassador to India, used language that he later recommended to his fellow chiefs of mission: "All ... communications ... go through me." This would ensure that "you know fully what is going on and can act as necessary against that of which you disapprove, including vetoing the transmission itself." Furthermore, telegraphy—unlike media capable of directly broadcasting to a large audience—operated from point to point. As a result, some people received information before others and became conduits for its diffusion.[10]

While the telegraph did not usually accelerate the pace of diplomatic life, one important exception exists—the decision to declare war, a decision that was too important to be left to foreign ministers. In such an instance, the telegraph could enormously hasten events by reducing the time necessary for the rulers of different states to communicate with one another.

By 1870, it was already observable that the telegraph had a huge impact on American diplomacy, especially through the use of the Atlantic cable. The telegraph, however, did have significant problems when being used in the area of diplomacy: expense, security, transmission errors, and the fear of using something new.

Expense was the first concern in the use of the telegraph. While messages sent overland were considered dear, the immense costs and lack of competition faced by the two transatlantic telegraph companies resulted in very high prices—more than five dollars per word in 1870. The State Department had good reason to be concerned about expenses since the diplomatic service of the United States was notoriously underfunded.[11]

The issue of telegraphic cost came about when the Franco-Prussian

2. Diplomacy

War began in 1870. Ambassador Washburne had such doubts about incurring the cost of telegraphing Washington with the news of the war that he sent the message by boat, also in the same message asking Secretary of State Hamilton Fish for guidance on when to use the transatlantic cable. This lack of warning due to cost left Washington unprepared to deal with the consequences of the conflict. The Franco-Prussian War depressed government bonds on which U.S. credit rested, disrupted trade flows, threatened diplomatic entanglement as both sides sought American support, led to breaches of U.S. neutrality by both belligerents, and created significant new diplomatic obligations (e.g., the American responsibility for protecting German citizens trapped in Paris). While the war did not threaten any vital U.S. interests, it did pose important problems for which American officials might have been better prepared had they received earlier warning.[12]

This desire to avoid spending money to send telegraphic messages led American diplomats to hope that journalists and businessmen would sent important political news at their own expense over the Atlantic cable to America.[13] This is a possible reason why the U.S. legation in Paris was remiss in reporting the coming of the Franco-Prussian War. When news of the French war declaration reached the U.S. government, it came via the Associated Press rather than by a governmental dispatch.[14]

This frugality continued to prevail over the timely reporting of important information. On July 13, 1814, Frank Mallet, the American vice consul general in Budapest, predicted war between Austria-Hungry and Serbia. But Mallet faced a dilemma; his superiors would have likely chastised him for sending this information to Washington. The telegram would have been about 150 words and Mallet knew that the State Department considered long telegrams extravagant and an excessive strain on those charged with deciphering them. Since this was the case, Mallet sent a letter by steamer rather than a telegram. His warning arrived fourteen days later, on July 27, 1914; Austria-Hungry declared war on Serbia the next morning.[15]

The telegraph's high prices resulted from the fact that their systems required skilled labor and an expensive infrastructure and, in turn, provided a scarce resource because, at first the lines could only carry one message at a time. The limited carrying capacity of telegraph lines allowed those who controlled the lines to ration access to them.

Austrian field telegraph station in Russian Poland, circa 1910–15 (Library of Congress).

Cost also became a factor in a different way. With the completion of the Atlantic cable in 1866, William Henry Seward, the U.S. secretary of state, sent a telegram whose high cost caused controversy. One reason that it was so expensive was that Seward, with an eye to increasing his popularity with the American electorate, sought to create the impression that he was capable of managing events from afar. He justified his long and expensive message by saying that he wanted his envoy to read it verbatim to Emperor Napoleon III. "For this reason," Seward noted, "no word was omitted under any consideration of economy."[16] Although Seward implied that the cable destroyed the autonomy of the U.S. minister to France, in the end it did not since the minister decided not to deliver Seward's telegram.

Another problem in the use of the telegraph by diplomats was that of security. Before the telegraph, foreign ministers gave their written messages to trusted couriers. Diplomatic pouches, envelopes, and elaborate wax seals provided additional protection from having their contents read. The telegram in this case offered less protection than giving

2. Diplomacy

the contents of the secret diplomatic dispatch to a telegraph clerk who was an unknown foreign national employed by a foreign government for transmission to one's country. The reason was that diplomats were forced to send telegrams through the state monopolies of their host countries and of those countries through which their messages travelled. This situation compelled foreign ministries to rely on encrypted messages in order to thwart the intelligence agencies of those countries through which the messages were sent. As a solution to the problem, beginning in 1867, the U.S. government gave its foreign ministries a cipher to be used in transmitting sensitive information over the Atlantic cable. Codes not only added security but also offered a means of shortening messages and saving money. In one historian's opinion, "the telegraph made cryptography what it is today."[17]

Garbled telegrams, caused by both human and mechanical errors, also limited the usefulness of the telegraph in diplomacy. This was not a problem when the message was sent unencrypted, because any mistakes were obvious. When encoded, however, errors became critical because, since they could not be read by the sending telegraph clerk, the mistakes went through. Additionally, since the encrypted messages were as brief as possible in order to save money, the error rate was far greater than it would have been using written diplomatic dispatches. After receiving a garbled message from the secretary of state on July 17, 1870, Hoffman noted, "I deciphered it with some difficulty, many letters in important words having been changed in the transmission." The same day, Hoffman received a second telegram from the same sender, which, "happily," was "not in cipher."[18]

Another garbled message from this period of time produced by human error in the system of diplomatic communications shows how serious these mistakes can be. The immediate causes of the Franco-Prussian War can be found in a mistake that the Prussian delegation in Madrid made when deciphering a telegram. The error, a substitution of the word "ninth" for the word "twenty-sixth," led to a delay in Bismarck's secret plan to place a Hohenzollern[19] on the throne of Spain. During the delay, news of the plot reached the French government, precipitating the crisis that resulted in the destruction of Napoleon III's regime and the creation of the Second Empire in Germany.[20]

The last problem involving the use of the telegraph in diplomacy was simply inertia. Even when one of its major drawbacks, the cost, ceased to be a factor, the diplomats were slow to use it. The cost of the Atlantic cable dropped at an average rate of 15 percent per year for the first twenty years after its successful implementation. In 1866, a telegram between Washington and London cost ten dollars per word; by 1904, the price dropped to twenty cents per word. Despite this large price reduction, there was not a corresponding increase in usage among diplomats.[21] This situation continued especially among American diplomats, who used the telegraph basically during emergencies until World War I. Before then the usual means of diplomatic communication was letters. The war became the impetus which pushed the diplomats into more frequent use of the telegram. During the war the State Department's expanded duties, larger budget, and more vigorous involvement in the competitive international arena induced it to utilize the telegraph with regularity. American diplomats began to employ it continuously. Although the State Department's use of the system declined after the peace negotiations, it remained higher than the prewar level.[22]

There was also the obvious possibility that the diplomats avoided the use of the telegraph because it lessened their autonomy. Labor history has frequently explored worker sabotage of technologies that threatened their independence on the job. While there was no evidence that diplomats purposely avoided using the telegraph to maintain their freedom of movement, they had obvious incentives to cover up instances of their insubordination.[23]

For another example of the use of the telegraph in diplomacy, it is useful to look at the British attempt to connect England with India via telegraphy. In this case the telegraph served two "empires," the British and the Ottoman. The reason was that in order to complete the landline between England and India, the British had to run the line the full 1800 miles through the Ottoman domains in Asia. For the British the telegraph acted not only to unite their empire, but also to enhance their political and commercial interests in the East. The Ottomans, on the other hand, exploited the advantages of electric communications to consolidate the control of their own empire.[24]

Samuel Morse made two attempts to sell the telegraph to the Ottoman court, first in 1839 and then successfully in 1847. However, it was not until the Crimean War that the first line was actually built. The basic

2. Diplomacy

problem was the pashas.[25] Cyrus Hamlin, then a missionary, later the president of Robert College in Istanbul, observed in the Ottoman empire what was a probably the case in Europe, the question of autonomy. Here the pashas had united against the establishment of the telegraph: "They wanted no such tell-tale to report their doings every day, while in the distant interior."[26] Even when the telegraph was being tested in the sultan's court, the wires had been found to have been mysteriously cut, possible by a pasha who wished the trial to fail. Since the pashas ruled arbitrarily in the distant provinces of the empire, the introduction of the telegram was not in their interests. With the telegraph, the sultan's orders could be quickly conveyed to the governors and officials, who could be summoned to Istanbul or be replaced without warning. Furthermore, public complaints and petitions about the pashas could be communicated to the sultan directly.

With the outbreak of the Crimean War in 1853, Britain and France joined the empire in an effort to prevent Russia from expanding its territory into the Mediterranean. This alliance allowed the Ottomans to put aside their opposition to new innovations which would have prevented the acquisition of the telegraph, usually based on religious grounds, since religious law held that innovations useful to winning a war were justifiable.

With the conclusion of the Crimean War, the Ottoman empire joined Britain and France in the construction of telegraph lines across their territory. Even though the Ottoman interest in telegraphy had grown substantially since the beginning of the war, they needed the know-how that the British and French possessed and those nations needed the empire to provide poles, labor, and protection for the completed lines.

When it came to a contract between the three countries, however, the Ottomans, influenced by the French, opposed the whole design of the telegraph line because their capital, Baghdad, was not chosen as the terminus of the line. Such dissension was an ordinary occurrence: Britain and France often competed for concessions in the Ottoman empire, each hoping to expand its sphere of influence through these undertakings, while the empire sought a tactical balance between them.

In this case, however, the empire went further. Increasingly aware of the economic and political implications of telegraph links among their major towns and provinces and Baghdad, they stopped all further negotiations on any telegraph systems controlled by the British or the French. The Ottoman officials insisted on keeping "the telegraph towards India in their own hands."[27]

The empire pushed its advantage as far as possible. While keeping overall control of the project, it agreed to employ British engineers and workmen that had been employed by the European and Indian Junction Telegraph Company[28] and to use the company's stock of supplies. It also offered to build the line with two wires, one to be retained by the Ottoman government for its own service, the other to connect England with India. Diplomatically, Ali Pasha, the Ottoman foreign minister, explained the dual advantages of this project; on the one hand it would fulfill Britain's aim of direct communication with India, while at the same time it would render an immense service to the provinces of the Ottoman empire.

The British and Ottoman diplomats worked out an agreement. Not only would the line connect England to India, but would also be a basic way to introduce "Western civilization" to the East.[29] The British also agreed to pay the Ottoman government an annual subsidy to use the line when it was completed. The Ottomans were required to purchase all equipment necessary to build the line from English suppliers and to furnish the needed laborers. The planners, construction engineers, surveyors, linemen, and medical officers were supplied by the British.

The line was completed in January 1865, allowing the first uninterrupted telegraphic communication between India and Europe.

While the telegraph's impact on the relations between the United States and France and also on its construction the Ottoman empire during the nineteenth century are two examples of the telegraph's effect on diplomacy, consider these two quotes to bring into perspective not only immediacy, but also distance. First from a Tennessee congressman during the 1790s "Situated as the United States are, at a great distance from the transatlantic world … what have we to do with the politics of Europe?"[30] Next, this assertion from a former American foreign diplomat, a hundred years later speaking of the telegraph, that it was among the new forces that "have brought foreign countries to our door and have carried us to theirs."[31]

Distance is what basically created foreign diplomats, since the head

2. Diplomacy

of a nation could not simultaneously rule at home and negotiate in a foreign country. This distance allowed the diplomat not only to convey messages back and forth between his host nation and his homeland but also to make decisions in the absence of the ruler of his country. Viscount Castlereagh, who had attempted to improve relations between Great Britain and the United States by revoking the "Orders in Council," noted, "The remoteness of America from the events passing in Europe mak[es] it of the utmost importance to the cultivation of a good understanding that an accredited minister, fully authorized to act for them in the many delicate cases that necessarily grow out of the present state of Europe and measures adopted by the Belligerents should be resident at this court."[32]

Even when events overseas touched upon vital national interests, distance compelled governments to devolve power to diplomatic envoys. The mission of James Monroe to buy New Orleans from Napoleon is illustrative. As President Thomas Jefferson said, no instructions could be "squared to fit" the circumstances that Monroe would be facing. Likewise during the crisis that preceded the Crimean War, Britain's foreign secretary gave the British ambassador in Constantinople complete authority to declare war:

> No later intelligence having since arrived, Her Majesty's Government are uninformed whether the Sultan ... has actually declared war.... [It] is difficult to send you precise instructions upon a state of things so critical, but with respect to which so much uncertainty still prevails.
>
> Her Majesty's Government are far from intending to impose upon your Excellency any responsibility that they can properly take upon themselves, but having explained the reasons which render precise instructions impossible, they think it advisable to leave your Excellency and to Admiral Dundas, in concert with the French Ambassador and Admiral, to determine upon the best mode of giving effect to their views for the defense of the Ottoman dominions against direct aggression.[33]

After completion of the telegraph line to India, colonial officials believed that they would now be able to better supervise their foreign proconsuls. Although the connection of England and India by telegraph enabled them to exercise much more power over England's provinces than they formerly had, the reverse was true in these provinces. British writer George Orwell, who had spent time as a colonial official in Burma, remarked that telegraphy reduced "the one-time empire builders ... to the status of clerks, buried deeper and deeper under mounds of paper and red tape."[34]

During the time before telegraphic communication, governments wrote dispatches on the basis of old information and realized that their instructions would be even more dated when they finally reached their recipients. In the interim, many problems would have solved themselves. At times this made the entire exercise of writing instructions seem pointless, especially in a country that had few foreign interests. Thomas Jefferson, when serving as secretary of state under George Washington, is said to have observed that it was over two years since he had heard from the U.S. minister to Spain. If another year went by with no news, he planned to write to him.[35]

The use of the telegraph, along with giving a nation better contact with its envoys, allowed the home government to claim the opposite, that there was no contact with an envoy, that the envoy acted on his own. This happened when governments needed to retreat from some position. This is known in diplomatic circles as "plausible deniability." In such cases the foreign minister takes the blame to protect their country when a policy goes wrong.

One such case involves President James Polk during the Oregon border dispute. When he was campaigning for office, he implied that if elected he would go to war against Great Britain unless the United States received the entire Oregon territory up to the fifty-fourth parallel (hence the famous slogan "Fifty-four forty or fight"). Once in office, however, Polk began to back off on his campaign pledge and go to the forty-ninth parallel when the negotiations stalled. Publicly, however, he still maintained that he would go to war if the British did not back down.

Louis McLane, the U.S. minister to Great Britain, followed Polk's actual instructions and worked out an agreement with the British putting the line at the forty-ninth parallel. Polk sent the agreement to the Senate without supporting it, which created a feud between Polk and McLane. Regardless of who was more correct, Polk's criticism of an agreement that his own agent had secured gained credibility because there were no written instructions that a telegraph would have provided.[36]

Perhaps the greatest effect that the telegraph had on diplomacy was the loss of identity in the person of the foreign minister. At one time chiefs of mission exercised enormous individual power as actual policy

2. Diplomacy

makers; now they had become cogs in the machine. Consider this article from the *New York Times* in 1900:

> It is a fact that [the diplomat] has become less of a statesman and more of a correspondent, an exponent of his master's views, a go between, an instrument. He is allowed little individual action these days. Every move in his negotiations must be sanctioned by the Minister of Foreign Affairs, at home, every objection or demand coming from the other side is referred to this same august autocrat. There are no Kaunitzers or Talleyrands in these times of the fast mail and the telegraph, when instructions more particular can be so easily asked, and supplementary orders so quickly transmitted.[37]

The effect of telegraphy on the individual diplomat varied greatly. In some cases, diplomats welcomed instantaneous communication with their foreign offices since it relieved them of the responsibility of on the spot decision making. Many diplomats, however, did not. A British envoy commented upon "this age of rapid communication, of what I would call the telegraphic demoralization of those who formerly had to act for themselves and are now content to be at the end of the wire." A British colonial agent groused, "I might have been a great man, but for the telegraph." Francis Bertie, who worked both in the British Foreign Office and as a diplomat, also complained, "In Downing Street one can at least pull the wires, whereas an Ambassador is only a d ... d marionette." Many diplomats felt that electric telegraphy deprived their activities of value,[38] some to the point that they refused a position. In 1913 President Woodrow Wilson offered Richard Olney the post of ambassador to Great Britain. Olney declined the offer, partly because his experience as secretary of state during the Cleveland administration had convinced him that "the English Ambassadorship is a show place—all the real work being done in Washington." Olney also noted: "An ambassador is nobody in these days. He sits at the end of a cable and does what he is told."[39]

Another use of the diplomatic telegraph in giving a nation's rulers a choice was whether to deal directly with a foreign country or use their envoy. Consider this from a British diplomat about the effect of the telegraph on this choice:

> When a crises of a certain magnitude occur, Foreign Ministers ignore their Ambassadors and correspond over their heads direct with one another.... The system is hard on the Ambassadors, diminishes their stature and destroys the relics of such of their initiative as has been left them by the

telegraphic tyranny under which they habitually live. It may, indeed, be necessary in order to gain great ends. But if it is, it only goes to prove that in grave issues the diplomatic machinery to which we have been so long accustomed, is antiquated, and, given the modern facilities for quick communications between actual rulers, ineffective.[40]

Two factors—crisis induced stress and the desire to act quickly—made political leaders likely to circumvent theirs envoys if they believed that circumstances warranted that course of action. The desire to deal directly with a foreign nation predated the telegraph; now, however, they could communicate in real time.[41]

Finley Peter Dunne's Mr. Dooley also had an opinion on diplomats and the telegraph: "In odher to keep up with what is goin' on th' Ambassadure has to get out arly an' buy th' mornin' papers. Th' first thing he knows about a war that's broke out between th' two counthries is whin he goes around to call on th' King an' a senthry jabs him with a baynit."[42]

Since the use of telegraphy in diplomacy tended to concentrate power in the hands of a nation's bureaucracy and to a lesser extent in the control of those who had the power before the telegraph's existence, its use became something to be avoided by some diplomats. Three examples of legates resisting the constraints of the telegraph are the British envoy to Constantinople, Lord Stratford de Redcliffe; the French ambassador to St. Petersburg, Maurice Paleologue; and Baron Wilhelm von Schoen, Germany's ambassador to Paris.

Lord Stratford de Redcliffe was described as "the last diplomat to hold the issue of war or peace in his hands." He believed the telegraphic communications to be "subject ... by their very nature, to the risk of erroneous information, or premature instructions." Also, telegrams were "liable to frequent mistakes in the transmission" and were sometimes so concise (because of cost) that diplomats were forced to rely upon their own judgment.[43] Since this was the case, it was Stratford's opinion that a diplomat was forced to use "superior judgment" (such as his own) to decide whether or not a particular message from the foreign office was to be heeded.

Some diplomats agreed with Stratford that the cost factor prevented London from micromanaging foreign affairs. In 1861, Lord John Russell averred that diplomats had more autonomy because of short, concise telegrams than they would have had from long, detailed written instruc-

2. Diplomacy

tions. Sir John Tilley, chief clerk of the Foreign Office during the First World War, agreed: "It is certainly the case that an Ambassador may be a good deal at sea as to the reason for his instructions and the spirit in which they are issued, and yet may not feel that he can delay action till he has obtained clearer explanation."[44] The basic problem was that telegrams carried small amounts of information very quickly, as against a letter which carried all the information slowly. The onus was on the countries' Foreign Office; if the telegram stated the intention of the message clearly, all was well, if not, the power of decision went to the diplomat.

Maurice Paleologue was another example of a diplomat who wished to act on his own, ignoring instructions from the foreign ministry. In 1914 he was appointed ambassador to St. Petersburg. His problem was that he had a great admiration for Russia and at the same time was deeply pessimistic about Germany's intentions. Because he believed that Germany would attack France before the Russians could mobilize,[45] he prodded the Russians to attack Germany.

What transpired subsequently became a prime example of a self-reliant diplomat who believed in the "great man" theory of history even in the age of instant communication brought about by the telegraph. Acting independently, Paleologue assured Russia that it could count on unqualified French support in any pledges that it made to Serbia. Secondly, he did not inform Paris of Russian mobilization. His motives were to ensure that the war which he saw as inevitable should come sooner rather than later, so that the French and Russian armies would not be at a disadvantage fighting the Germans without the support of the other.

On July 31, 1914, French Prime Minister Rene Viviani, unaware of Paleologue's actions and attempting to avoid war, sent a telegram to Paleologue which ended, "As I have already informed you, I have no doubt that the Imperial [Russian] government, in the greater interest of peace, will avoid, for its part, anything which could begin the crisis." However, by this time Russian mobilization had already begun and war was inevitable. Because of their ignorance of events, the French leaders had no opportunity to influence the decisions made by Russia, decisions that would have far-reaching effects on their own country.[46]

On August 3, 1914, Germany's ambassador to France, Baron Wilhelm von Schoen, presented a German declaration of war to the French prime minister. The German government had instructed Schoen to tell the French that the reason for the declaration was that French aviators had committed hostile acts over German territory and that French ground troops had crossed the border into Germany. Schoen only discussed the air strikes with the French, claiming that the part of the telegram that discussed the border incursions was garbled. Schoen later wrote that, "the telegram was so mutilated that, in spite of every effort, only fragments of it could be deciphered." He added, "As I knew from other sources that we felt bound to declare war in consequence of a French air attack on Nuremberg, I had to make up my mind to fall back on the little that could be clearly understood from the telegram, to justify the declaration of war."[47]

Why had Schoen falsely claimed that the telegram conveying the German declaration of war had been garbled? Was it to spare himself the shame of presenting a dishonorable war declaration? In any case, Schoen's behavior, like that of Stratford and Paleologue, demonstrates a tactic that diplomats could use to escape oversight from their foreign offices, because of the use of the telegraph.[48]

During the nineteenth century, most nations attempted to concentrate decision making at the top and the telegraph fit into this reasoning. Centralization through the use of the telegraph, however, had a downside, which was pointed out by a historian of the British Colonial Office: "Questions great and small, which formerly were decided on the spot, were now thrown upon the Home Government, not with the full information of a dispatch, but with the elliptical curtness of cipher telegrams."[49]

The drafting and revision of telegrams took time. Diplomats tended to choose every word in a telegram with great care because of the need for brevity (due to cost concerns) and the dangers of ambiguity. Through the period of the First World War, foreign ministry clerks manually coded and decoded messages, a laborious process that could significantly delay the transmission of messages. In the case of a message of particular importance, a foreign ministry official might even cable a warning to a distant diplomat, such as, "You will receive a very important message as soon as it can be put into cipher." The seriousness of cipher related holdups led foreign ministries to send uncoded messages when speed was important and secrecy was not.[50]

2. Diplomacy

Other problems impeded speed in the sending of messages. First was the fact that if the message was travelling a long distance, it would have to be retransmitted at various relay stations in order to boost the signal. As an example, a cablegram from London to Australia in 1900 took six hours to complete the distance. Also, telegraph companies often repeated encoded telegrams, since such messages were incomprehensible to the telegraphers and therefore prone to garbling. They might also recapitulate transmissions when they had reason to believe that an error had occurred. The manager of a telegraph company warned the German embassy in Washington, "Due allowance should be made for the time required for transmission and repetition." Even as late as 1998, when the British Foreign Office announced that it would stop using telegraphy, newspapers noted that a telegram not marked "urgent" typically required twenty-four hours to reach the intended minister.[51]

Of all the diplomatic telegrams that have ever been delivered late, the most famous is probably the final message from the Japanese government to the American secretary of state, Cordell Hull, breaking off negations before the attack on Pearl Harbor on December 7, 1941. Because the telegram reached Hull the day after the attack, it became a symbol of Japanese perfidy and stimulated the American war effort.

While it is not known precisely why the telegram was delivered late, there were a number of factors involved: first, the telegram was not marked urgent; second, the Japanese used decoding machines to decipher secret messages and two of their machines had been destroyed, slowing the decoding process; and third, the message arrived with missing and garbled words. In the age of the telegraph, communication delays were always possible and sometimes probable.

Even the perception of the telegraph's speed caused problems for diplomats. Consider these words in 1864 from Charles Francis Adams, the American envoy to Britain:

> Although entirely friendly to [the cable] scheme, I must confess I am not very anxious it should be carried out immediately. It is a very great object no doubt to bring the two countries together, but I cannot help arguing with myself that if, with the two countries three thousand miles apart, I get so many dispatches per week that I can with difficulty attend to them all satisfactorily, what would be my fate if the cable succeeds, and I had to receive and answer them every day? Therefore, I shall wish success to the Submarine Telegraph between Europe and America, but [may it] happen with just about

as little delay as may bring it to the moment when I hope to be back in my native country.[52]

Another diplomat, Edmund Hammond, the permanent undersecretary at the British Foreign Office, spoke of another consequence of the telegraph's speed: "I dislike the telegraph very much. In the first place nothing is explained by it, it tempts hasty decision." Concise telegrams often conveyed enough data to induce panic without imparting the information necessary to provide understanding.[53] An editorial in the London *Spectator* made the same point:

> It is rumour rather than intelligence which is hurried so breathlessly across continents and seas.... The constant diffusion of statements in snippets, the constant excitements of feeling unjustified by fact, the constant formation of hasty or erroneous opinions, must in the end, one would think, deteriorate the intelligence of all to whom the telegraph appeals.... Is it conceivable that statesmen, informed of all they know through such a medium, with its inevitable hurries, its inherent necessity for over-compression, its unavoidable reticence's, should ever know really *know* even events, should ever be wisely counselled, should ever be internally urged to reflection, as they were under the old *regime*?[54]

The use of the diplomatic telegram gave rise to a new form of espionage—code breaking. Instead of relying entirely on spies, countries also attempted to collect information from the messages being sent from the field to the home ministries. The ease with which a code breaker could read an encoded message depended on the code maker. This was a difficult and time consuming endeavor because the cipher had to be: cost-effective, concise, secretive, reliable, easy to use, and yet difficult to break. If this was not the case then a diplomatic telegraph message became, in the words of the secretary of the French Atlantic Telegraph Company, an "open letter." This was especially true when a nation's telegraph system had been nationalized, which was the case in every major power except the United States. As a result, diplomats in foreign countries were wise to assume that their telegraphic messages were being read by their host's intelligence services. Thomas F. Bayard, the U.S. secretary of state, warned, "It is generally understood that some governments, especially those whose telegraphs are under the charge of the central administration, spare no effort to decipher the secret messages of foreign representatives." A former Austrian code officer, describing the years before the First World War, noted, "Our telegraphers, like those

2. Diplomacy

of other governments, were instructed to send us a copy of every cipher dispatch that passed over their wires." During the First World War, the Vatican was the most striking example of a diplomatic power vulnerable to intercepted messages: once Italy entered the war, at least one of the belligerents, and frequently more, could obtain every telegram from the papacy.[55]

Looking at diplomatic telegraphy from the viewpoint of efficiency, there is no doubt that there were many advantages. Faster communication which facilitated day-to-day diplomacy allowed rapid answers to incoming questions and timely solutions to sudden problems. The telegraph offered unprecedented capabilities to its users, as remarkable for its time as those afforded in recent years by other new means of conveying information such as the Internet. It promised the nearly instantaneous transmission of complex messages over enormous distances, freeing information from dependence on human carriers. Even more than the steam engine, transoceanic telegraphy marked a liberation from the influence of weather conditions that imposed so much uncertainty on communication by sailing ship. Yet diplomats, from code clerks to ambassadors, did not always feel that they had benefited by its use. Many complained about a technology that reduced their authority, lessened their control over their work environment, and degraded their working conditions.

Nevertheless, this revolutionary technology did not produce a revolution. The Atlantic cable had in fact gradual consequences for diplomacy, since, despite the influence of telegraphy, ship-carried diplomatic pouches continued to play a predominant role in American foreign policy until the First World War.

Moreover, as the historian Svante Lindqvist suggests, older technological systems embody values that influence "the thoughts and social actions" of later periods.[56] This may explain the lag between the creation of a technical capability, in this case the telegraph, and its acceptance by foreign ministries. Because of this disconnect, the United States and other nations maintained diplomatic practices no longer adequate in a rapidly changing world; in other words, their adaption to the telegraph lagged dangerously behind their adoption of it.[57]

3

The Telegraph and the American Civil War

By the time of the American Civil War, several European nations had made some military use of the electric telegraph, most notably in the Crimean War. How well this European experience was known in the United States is not evident, but in any case, the American Civil War, which was first in many things, led in the extensive employment of the telegraph for wartime purposes.

The military telegraph was simply an expedient. News of the attack on Fort Sumter on April 12, 1861, was carried throughout the nation on the three principal private telegraph companies—the American, Western Union, and Southwestern. On February 26, 1862, under legislation passed during the preceding month, the president took control of all telegraph lines in the United States, which meant in practice that the United States Military Telegraph (USMT) could use all American telegraphic services as circumstances demanded.[1] The USMT was not a true military organization, but a civilian bureau acting under orders from the secretary of war. The operators were civilians and remained that way so that they would be immune to orders from army officers. Many of the telegraphers were teenagers, and some were women.[2]

The USMT was nominally attached to the Quartermaster's Department. In reality it was responsible directly to Secretary of War Edwin Stanton. This arrangement caused a great deal of friction on the front lines between the operators and the military commanders. This friction resulted from the fact the operators, showing abundant common sense, usually tried to accommodate the army commanders in any request that was not contrary to their orders from Washington; however, since much of the telegraphers' messages were in cipher, it is easy to see why the commanders believed that they were not being given all of the available

3. The Telegraph and the American Civil War

military information.[3] Just as the military personnel, however, the telegraphers were required to take an oath: "You do each and all swear to be true and faithful to the United States of America; that you will faithfully transmit all orders and information touching the military operations of the United States of America truthfully, and further, not to transmit any information, directly or indirectly to the Confederate States, or any representative or individual of said States, and in no wise prevaricate any information touching the military operations of the United States of America."[4]

During the Civil War, the army's signal corps commandeered innumerable high observation posts and signaled by flag and torch (at night)[5] from hills, high buildings, specially constructed towers, tree platforms, and even the masthead of ships. But signals by flag and torch had the obvious weaknesses of all visual signals; they were dependent upon line-of-sight and they were obscured by bad weather and by the smoke and fog of battle. When it was given its first real test in war, the electric telegraph rendered the signal method old-fashioned.[6]

On April 19, 1862, all telegraphic communication between Washington and Richmond, Virginia, were cut. The telegraph companies in the North not only continued to operate, but, because of their patriotism, there was never any complaints from their stockholders because of the government takeover of their businesses. It was seldom required for the Union forces to actually possess any telegraphic office in the North. For the most part, military control of private lines was merely nominal—not for want of power and authority, but because of the ardent cooperation of the companies' personnel. Nevertheless, there were instances where military officers, especially those on the staff of General George McClellan, objected to the use of the telegraph, perhaps because they believed that it would not be useful in battle[7]; but since McClellan had faith in its ultimate efficiency, it was used throughout his command beginning with the Peninsula Campaign. In order to move with the army, the telegraphic instruments which comprised a power source and a sending unit were housed in a brass bound walnut chest. The box was then placed on a light-wheeled horse drawn covered wagon. The telegraphers with his forces kept McClellan fully advised as to all operations on his front. To McClellan, it was worth regiments of soldiers to *know* of an attack at any point along his lines, which were many miles long, so that the enemy advance could be met. As Union forces advanced into West Virginia in July 1861, McClellan wrote: "The first field telegraph that ever advanced with an army in America kept pace with this one."[8]

In December 1862, the field telegraph received a major test, this time at Fredericksburg. For the first time, the telegraph was used extensively in a tactical way on the battlefield. The lines were extended from the Union headquarters to the ends of both wings of the formation. The telegraph performed well, allowing the Union command to monitor events along the battle line. Some problems occurred when Union soldiers cut the lines thinking them to be a Confederate device. After instructions were passed down the chain of command not to interfere with the telegraph, the problem ceased.

Even though the telegraph had done well at Fredericksburg, significant problems emerged with the generators. One common malfunction was the lack of synchronization between the sending and receiving units, causing the receiver to bring in the wrong letter. An even greater problem was lack of range. The signal generated was fairly weak, and unintelligible after five miles.

The problem of lack of range became manifest during the next major battle at Chancellorsville. Here, the army was spread over a much greater distance than at Fredericksburg and the generators could not carry messages over the entire field. For crucial hours, from the evening of April 30 to the afternoon of May 1, the failure of the telegraph system left the right wing of Major General Joseph Hooker's army with no means of communication.[9]

During the war, the military telegraph expanded continuously in the North, adding one thousand, one hundred and thirty seven miles of line, on which were built one hundred and six offices, worked by one hundred and sixty three operators. In addition to the operators, the USMT also employed line builders, a category that included foremen, wagon makers, teamsters, messengers, battery keepers, linemen, watchmen, and general laborers. During the course of the war, about three hundred operators fell as casualties to wounds, disease, or capture. Repair of the lines was sometimes conducted under fire in battle and more often in country infested with southern sympathizers. Conditions were so bad in Tennessee that the linemen only worked under heavy escort.

3. The Telegraph and the American Civil War

The USMT used existing commercial lines throughout the North for much of its operation and constructed or repaired similar lines as Union forces retook Confederate territory. In his message to the army on November 14, Anson Stager, the manager of General McClellan's military telegraphs, reported:

> In many instances the wires followed the march of the army at the rate of eight to twelve miles per day, there being no other lines of communication upon the routes where these lines have been placed. The capacity of the telegraph for military service has been tested, and in affording rapid communication between the War Department, the Commander-in-Chief and the different divisions of the army; in directing the movement of troops and the transportation of supplies, it may be safely asserted, that it is an indispensable auxiliary in military operations. The organization of the government Telegraph Department, under the direction of the Secretary of War, will add greatly to the efficiency of this branch of the service.[10]

Stager, because of his work in organizing the military telegraph system, was appointed assistant quartermaster of volunteers in November of 1861 and commissioned colonel in February of 1862. Thus organized, the USMT became the medium of communication by which hundreds of armed men were directed from point to point, commissary, substance and ordnance stores ordered, and the innumerable necessities of great armies made known, as well as the reports of their victories and defeats.

But as highly developed as the telegraph was, its major flaw during the Civil War was that it was not controlled by the military, but rather a separate command. Anson Stager was responsible only to the secretary of war. He had complete control over the telegraphers who cooperated with the army only when directed, as an example: "*All Operators in the Department of the Tennessee:* Mr. J.C. Van Duzer has been assigned to the management of the United States military telegraph lines in the Department of the Tennessee. You will obey instructions received from him. Orders from any other source will not be obeyed. Anson Stager, Colonel and General Superintendent of Military Telegraphs."[11]

The result of the independence of the telegraphers from the army and department commanders sometimes rendered the military telegraph useless and on occasion a menace. More than once in the Civil War important operations did not come to fruition because instead of putting one commander over the other, thus making one responsible, they were asked to cooperate.

How the Telegraph Changed the World

An example of this occurred on the first day of the battle of Spotsylvania, as reported by Union General James Wilson. The Union Fifth Corps under General Gouverneur Warren had been slowly pushing back the Confederates who were blocking the Union advance to the Spotsylvania Court House. Next, two actions happened simultaneously, the Fifth Corps was joined by the Union Sixth Corps commanded by General John Sedgwick and the Confederates sent in reinforcements. Now with both sides having an equal force, a stalemate ensued. General George Meade came on the field and seeing that independent action on the part of the two corps under his command were achieving nothing, sought to bring about united action. As General Wilson tells the story:

> Meade turned to Warren and said: "Warren, I want you to cooperate with Sedgwick and see what can be done." Whereupon Warren, who had been Meade's Chief of Staff and had doubtless been accustomed to speaking plainly with him said: "General Meade, I'll be God damned if I'll cooperate with Sedgwick or anybody else. You are commander of this army and can give your orders and I will obey them; or you can put Sedgwick in command and he can give the orders and I will obey them; or you can put me in command and I will give the orders and Sedgwick shall obey them, but I'll be God damned if I'll cooperate with General Sedgwick or anybody else."[12]

If major generals commanding an army corps, when asked to cooperate, could not, what was the chance of lower level individuals, one group in the military, one not, cooperating? And what were the consequences if they did not? Especially when they did not have a common superior present. The answer was in many cases a lack of cooperation that led to lost opportunities, and at its most serious, the loss of a battle.

Since the telegraph department operatives were not part of the military chain of command, there developed a spirit of insubordination, of independent judgment on the part of each telegrapher as to where he should go and what he should do. Colonel Stager had developed an *esprit de corps* among the telegraphers that could have been a good thing but unfortunately, its practical application was to have the telegraphers unite against the army commanders whenever pressure was brought against any one of them.

In November 1864, Major General John Schofield's command was engaging the Confederates as he moved toward Nashville, Tennessee,

3. The Telegraph and the American Civil War

when he was deprived of the use of his telegraph because of his three operators. One was arrested for leaving his post, another had gotten drunk, and the third would not return to work unless the first two were released. J.C. Van Duzer, in charge of the telegraph operators in the Department of the Tennessee, had the men reinstated and additionally contacted Colonel Stager criticizing General Schofield. General Grant also had a problem with Van Duzer. Grant wished to have the telegraph line extended to his commands' front as the army advanced. Van Duzer wired Stager that the operators wished to leave Grant's command rather than comply and asked that Grant's men lay the wires. Grant relieved Van Duzer of his command so completely that Stager could not reinstate him at that time. These are examples of a situation in which a military telegrapher exercises his independent judgment as to whether or not he should cooperate with an army general.[13]

Grant also had a problem with the telegraph after his victory in the taking of Fort Donelson in Tennessee. At that point in the war, Grant was attached to Major General Halleck's command. When Halleck did not hear from Grant for a week after the victory, he contacted McClellan in Washington on March 3, 1862, reporting that Grant was doing nothing, that he was not in the field pursuing the Confederates. In reply McClellan (with the authority of the secretary of war) telegraphed Halleck to arrest Grant and give his command to General Charles F. Smith. When it was brought to his attention that Halleck had not received any of his telegraphic reports, Grant claimed that he had sent telegrams to Halleck's adjutant general every day after the battle. On March 15, Halleck withdrew his charges and explained that the failure to receive Grant's reports were due to what he called "interruption of telegraphic communication." The possible reasons for the "interruption" was that Halleck wished to supersede Grant and also that President Abraham Lincoln called for a bill of particulars from Halleck concerning Grant's actions, in which case Halleck would have had to prove specific charges. In this instance the telegraph proved to be a convenient scapegoat.

A military telegrapher attached to an army engaged in battle and wishing to avoid its hardships often simply left his post. When Grant moved against Vicksburg, Van Duzer, still with him at the time, insisted in staying behind in Cairo, Illinois. In his place, Grant had to supervise the extension of the telegraph lines behind the advance and also set up

How the Telegraph Changed the World

the telegraphic offices. Again, on one occasion during the battle of the Wilderness in 1864, there was a delay of twelve hours in Grant's dispatches to the War Department because the telegraph operator could not be found. This incident brought a sharp letter from Secretary Stanton to Grant's superintendent of military telegraphs. The superintendent's reply was that he had a work party erecting the telegraph poles and expected to have direct communication to Washington the next day. As Grant had moved ten days before, it is evident that the telegraphers were in no hurry to extend the front until the secretary of war intervened.

Often the operators were simply cowards and would not stay where they were needed, but ran off to a safe position, where they were of no use. If ordinary soldiers had done this they would have been severely punished, but attempts of military commanders in the field to punish or even discipline these offenders were regularly thwarted by Washington.

There was a particularly flagrant example of this type of dereliction of duty on the part of the telegraphers at Gettysburg. During the fighting, it took twenty-four hours for dispatches to reach Washington from General Meade's headquarters. Even worse, on the morning of July 3, Meade wired General Halleck that the transmissions received the previous day from Washington were undecipherable because the telegraph operator, named Caldwell, had disappeared with the code. On July 4, Meade again wired Halleck that he was still unable to read the dispatches, now forty-eight hours old; additionally, Meade had found out that Caldwell had insisted on his independence and had moved the army's telegraph office to Westminster, Maryland, twenty-five miles away. It seems that he had become frightened during the fighting on the second day, which had gotten uncomfortably close to Meade's headquarters. Now at Westminster, he was safe, but useless.[14] This was the same situation that Grant faced at Vicksburg, with the telegrapher leaving the scene of the battle. To solve the problem in this instance, Brigadier General Herman Haupt, the superintendent of railroads, set up a horseback messenger service between Gettysburg and Caldwell in Westminster so that Meade's dispatches could be sent to Washington. To make matters worse, Meade ordered Caldwell's "immediate presence" at Gettysburg and the order was ignored.[15]

In July of 1864, Confederate General Jubal Early moved North

3. The Telegraph and the American Civil War

through the Shenandoah Valley towards Washington in an effort to capture the city and take the pressure off General Robert E. Lee. When he reached the Monocacy River, he was met by a smaller force commanded by Union General Lew Wallace. Although finally driven from the field by Early's overwhelming force of infantry, cavalry, and artillery, Wallace was able to hold Early back for two days so that his chance to take Washington was lost. Wallace's reinforcements were on the way to the battlefield by train which stopped thirteen miles short of the battlefield for want of orders. The reason for the lack of orders was because Wallace's telegraph operator not only ran from the battle, but also cut the telegraph line. As a result, General Wallace had no means of communication with the rest of his troops, nor they with him, although the troop trains were in immediate proximity to the telegraph line along the railway which ran directly to the bridge where Wallace had set up his headquarters. John Garrett, president of the Baltimore and Ohio Railroad, realizing the situation sent the trains to Wallace; however, it was too late to defeat the Confederates. The battle that Wallace had called "the best little battle of the war" could have been much more but for the lack of the telegraph.[16]

In this instance, as in most of the cases of operator maleficence that were disclosed by the records, the result was the same. The reports were received in Washington, placed on file, and ignored.

Probably because there was no action taken against incompetent and cowardly operators, these instances continued. The most flagrant case, and one which came very near to having fatal consequences, happened in the Franklin-Nashville campaign which began on September 18, 1864. The situation was as follows: Union General William T. Sherman was traveling west to the sea after leaving Union General George H. Thomas in control of Atlanta. Confederate General John Bell Hood was moving north into Tennessee with a large army; Union General John Schofield with his smaller force was in front of Hood with orders to hold him up until Union General Thomas could join him with his army. Hood was eager to attack Schofield since with his larger force, he would have been victorious. At this critical juncture, McReynolds, Schofield's telegraph operator, became scared. He left Columbia and set up his office in Franklin, Tennessee, well to the rear. Hence the telegraphic messages from General Thomas to General Schofield traveled from and to Franklin, Tennessee, not to General Schofield in Columbus, Tennessee. McReynolds then decoded the messages which were then sent on horseback to General Schofield. This system created two problems; first, the amount of time that the transmission of the message took, and secondly, because the messenger had to ride through enemy territory where he could be captured by a hostile population or by Confederate General Nathan Bedford Forrest's cavalry. In fact two such messages were intercepted.[17] Here again, it can be seen that the split command causes enormous problems. The commanding general's staff—the chief commissary officer, the chief quartermaster, the ordnance officer, even the chief of railroad transportation—were at his disposal, but not the chief of the telegraph unless he so wished.

Under pressure from Hood, Schofield began a slow retreat from Columbia to Nashville, and now the problems with the telegraph became manifest. On November 25, Thomas complained that it was taking twenty-four hours to send a telegraph from Nashville to Schofield in Columbia, a distance of forty miles. On November 27, Schofield wired Thomas that he was having trouble deciphering the telegraphed instructions after a twenty-four hour delay in receiving them because McReynolds was still in Franklin. On November 29, the telegraph operator at Thomas' headquarters reported that Thomas' cipher operator, Everett, had not as of 6 a.m. sent off messages given to him at 3:30 a.m. By this time it was taking up to forty-eight hours to get messages from Thomas to Schofield. The reason, as it turned out, was that Everett was a drunkard. Worse, the person in charge of the military telegraph at Thomas's headquarters was Captain Van Duzer, who had caused problems for Grant.

On November 30, Schofield had had enough. He wired Van Duzer that he had arrested McReynolds for disobedience of orders, Everett for drunkenness, and the third operator for swearing that he would not work until McReynolds and Everett were released. Van Duzer's only action was to complain to Major Eckert in Washington. Eckert, who hated Schofield, held up Grant's orders when it seemed that Thomas was reluctant to move against Hood.[18] McReynolds was tried by court-martial under Schofield's orders, convicted, and sentenced to a month's hard labor. By order of the War Department, the sentence was at once remitted and McReynolds was restored to duty.[19] In other words, the

3. The Telegraph and the American Civil War

War Department promptly made it clear that the military commander in the field had no power over the use or disuse of his telegraph. Fortunately for Schofield, even with the telegraphy problems, he managed through skill and good fortune to obtain his objective.

As has been seen, some of the operators in the military telegraphs administration were cowards, some were disinclined to face hardships, and some were drunkards; additionally, there was the profit motive in using the telegraph. Operators could make money by sending press dispatches and messages likely to affect the stock or cotton market. Therefore there was a strong temptation and in too many cases, a habitual tendency on the part of military telegraph operators, to give precedence, even in military emergencies, to business and press dispatches.[20]

This was probably the reason for the telegraph problems that Grant was having even when he was not in the midst of a battle. The general was injured after falling from his horse at Vicksburg. As soon as he was able to return to duty, he was ordered to go to Cairo, Illinois, and report to Washington upon arrival. He arrived on October 16, 1863, and at 9 a.m. reported by telegraph that he was ready to take the field. This message, which the War Department was eagerly awaiting, was not received until 10:40 p.m. At 11:30 that morning, Grant wired Washington again with some routine information, this message was received at 9 p.m.[21] In other words, the important message, sent to be telegraphed two and a half hours before the unimportant one, arrived an hour and forty minutes later. It is more than probable that the important message was laid aside until private and press messages had been sent out, and then when the military dispatches had piled up were sent in inverse order of their receipt.

Grant believed that the problem was the operators because he had seen this happen before. In 1862, when he was moving his army from Corinth to Vicksburg, he found that his dispatches were frequently delayed for an entire day. Discovering that the reason was that the operators were making extra money sending private messages, he instituted an order allowing non-military (i.e., cotton and press) dispatches to be sent out only before 10 a.m. and only if they did not interfere with military business. He also put one of his aides in charge to see that the order was carried out. When later Grant found that his orders had been ignored and his dispatches sent late, he asked the operator for an explanation.

How the Telegraph Changed the World

The reply was that the telegraph was used to send out cotton price information and that the operator's superior was the same Van Duzer that Grant had had problems with before. Grant sent the following report to General Halleck:

> I reminded Van Duzer that my order ... was still in force.... Mr. Van Duzer replied that my orders should be obeyed, but immediately removed the operator who had always been at my headquarters office and put in a new man, evidently because the first one had done his duty in informing me why my dispatches had been detained. I sent for Mr. Van Duzer and warned him against making changes at my headquarters without consulting me. I permitted the change to take place, however, notifying Mr. Van Duzer that I would have no person about the office who would not let me know when dispatches could not be sent and why.[22]

At this point, Grant began to make an issue of the problems that he was having with the telegraphers in general and Van Duzer in particular. The problem was that Van Duzer had control of communication by wire with Washington. To get rid of operators that would side with Grant, Van Duzer telegraphed the War Department that "a man signing himself John Riggin" was interfering with the management of the telegraph in Tennessee. This brought a message to Grant from the War Department ordering the arrest of Colonel Riggin as an imposter. Grant's explanation was of no value, since he was told peremptorily that "Colonel Stager has appointed deputies believed to be competent," and any dereliction should be reported to Washington, not dealt with on the spot.[23] In addition, an order which Grant believed came from Van Duzer went out to "all operators in the Department of the Tennessee" to obey no orders but those from Van Duzer. Grant then arrested him. In support of Van Duzer, Grant's operators threatened to go out on strike, but Grant told them that if they did or did not stay on the job, he would not only arrest them, but put them to work at hard labor.[24] Immediately, telegrams began arriving purporting to come from the War Department to post commanders all along the line, ordering that Van Duzer be released. Grant believed that these orders came from Van Duzer's associates, and so did not release him. Indeed, Grant when so far as to threaten to shoot him if he remained in his command, since his willful insubordination and obstinacy made him a menace.

Amazingly, this was not the end of Van Duzer's career. Colonel

3. The Telegraph and the American Civil War

Stager simply replaced him in Grant's command and sent him to Nashville, where he became superintendent of military telegraphs for the Department of the Cumberland.

Even more serious than the difficulties with the military telegraphers already mentioned was the operator's monopoly of the cipher. This was jealously guarded, and as has been seen, this control deprived Schofield of Thomas' orders and directions in the critical period between Columbia and the Battle of Franklin. Two more examples both again concern General Grant.

In June of 1863 as he was preparing to attack Vicksburg, Grant telegraphed Schofield in St. Louis (the message went to Memphis by steamer and from Memphis to St. Louis by wire) for additional supplies needed for the siege. The wire from Memphis to St. Louis was in cipher. When it arrived at Schofield's headquarters, the operator could not be found. Fortunately, General Schofield had a background in decoding ciphers; even so, it took him twelve hours to unravel the code and then send the supplies.[25]

Another incident involving Grant occurred in January of 1864. The general had travelled to Knoxville, Tennessee, from Nashville to oversee operations. While there, he continuously received messages from the War Department which could not be read because his telegraph operator who had the codes remained in Nashville. To solve the cipher problem, on his return to Nashville he summoned the operator and ordered him to give the code breaking information to Colonel Cyrus Comstock, his aide-de-camp, so that there would always be someone on hand to read messages. The operator reported this to Colonel Stager, who promptly removed the operator, put in a new one, and a new cipher. Grant's contacting the War Department about this incident was not only useless, but earned him a rebuke for taking the sensible course of action under the circumstances. It is not easy to find an answer to Grant's observation in his letter to Halleck: "I could see no reason why I was not as capable of selecting a proper person to entrust with this secret as Colonel Stager."[26]

General Jacob Cox, who commanded the Union Twenty-third Corps and saw the practical workings of the operator's exclusive control of the cipher, believed that Secretary of War Stanton wanted a system in which the operators controlled the cipher and he controlled the operators so that his military commanders could send no messages which he or the department subordinates could not read. While this is only a theory of the general's, the following story was told by General Horace Porter, as related by Senator James Nesmith to General Grant:

> One day the Secretary of War sent me a message that he would like to see me at the War Department, at the earliest moment, on a matter of public importance.... I hurried up to his office, and when I got in he closed the door, looked all around the room like a stage assassin to be sure that we were alone. Then thrust a telegram under my nose and cried, "Read that!" ... I ran my eye over the dispatch, seeing that it was addressed to me and signed by [General] Ingalls [chief quartermaster of the armies in the field] and read: "Klat-a-wani-ka sit-kum mo-litsh weght o-coke kon-amox lum." Stanton, who was glaring at me over the top of his spectacles ... now roared out, "You see I have discovered everything!" I handed back the dispatch, and said, "Well, if you've discovered everything, what do you want with me?" He cried: "I'm determined, at all hazards, to intercept every cipher dispatch from officers at the front to their friends in the North, to enable them to speculate in the stock-markets upon early information as to the movements of our armies." I said: "Well ... it seems to me you omitted one little matter: you forgot to read the dispatch." "How can I read your incomprehensible hieroglyphics?" he replied. "Hieroglyphics-thunder!" I said: "Why that's Chinook ... Chinook is the court language of the Northwestern Indian tribes. Ingalls and I, and all the fellows that served out in Oregon, picked up that jargon. Now I'll read it to you in English: 'Send me half barrel more that same whiskey.' You see, Ingalls always trusts my judgment on whiskey!" ... That was too much for the great War Secretary, and he broke out with a laugh ... but I learned afterward that he took the precaution, nevertheless, to show the dispatch to an army officer who had served in the Northwest, to get him to verify my translation.[27]

While, as I have described, there were many instances in which the telegraph was a problem during the Civil War rather than a benefit, if it was used correctly, it became a huge advantage. This happened when the operators were competent, when they did not use the wire for their own gain, and were patriotic. An example occurred on March 25, 1865, when Fort Stedman at Petersburg, Virginia, was under attack by Confederate forces. Generals Meade and Grant were at Grant's headquarters in City Point, Virginia, far from the scene. Using the telegraph, Union General Henry Jackson Hunt was able to bring up his artillery and communicate with General John Parke, who was now in charge of the Army of the Potomac. General Parke, also using the telegraph, ordered a divi-

3. The Telegraph and the American Civil War

A field telegraph, battery wagon and officers tent of Military Telegraph Corps., headquarters Army of the Potomac, 1864 (Library of Congress).

sion of the Sixth Corps coupled with two divisions of the Fifth Corps and his own Ninth Corps to begin a coordinated attack ending in a defeat of the Confederates. This movement of troops could not have been done as quickly and effectively by messengers on horseback.

Actually, the use of the telegraph was routinely praised by the military commanders in the North, in spite of its problems.

The Union commander who probably made more use of the telegraph during the war was General Ulysses S. Grant. In 1864, when both Grant and General George Meade were moving south, their military telegraph unit was laying wire at the rate of about two miles an hour. Where the lines ran through territory unoccupied by Union infantry, cavalry patrols monitored the cables to prevent sabotage. The local inhabitants were held responsible for the safety of the cable; anyone caught tampering with it was to be shot on the spot. Grant began using the telegraph for strategic troop deployment and to keep a rein on his subordinate commanders early in 1862. In 1864, Grant was in almost daily contact with Sherman's forces in Georgia as they joined in combined operations.[28] Grant attested to the industriousness and reliability of the members of the USMT in his memoirs. "No orders had ever been given to establish the telegraph. The moment troops were in position to go into camp, the men would put up their wires."[29] The advantages of the telegraph over transmission of orders by aides on horseback is manifest. If, for example, General John Pope could have sent by wire the orders which went by horseback at Second Bull Run or General Robert E. Lee could have used the telegraph to make the attacks on the second day at Gettysburg simultaneous, the results in each case might have been different.

After the war the Union commanders were effusive in their praise of the telegraph and the operators:

Head-quarters Military Division of the Missouri
Chicago, September, 24 1879
W.R. Plum, Esq., Chicago, Ills.:

My Dear Sir:—Your letter of the thirteenth instant is at hand, and I have carefully considered its contents. I fully agree with you that the duties assigned to the Military Telegraph operators during the war, were, as a general thing, well performed, and the men were almost universally trustworthy. In my own experience, I found them invariably active, brave, and honorable; and I am glad to know that someone who is cognizant of their work, and who can speak from personal experience, proposes to write their military history. I have no doubt, but that individual acts of heroism can be found, that will show them to the public in a new and surprising light, and place them, and their work, on a higher plane than they hitherto have held in the estimation of the loyal citizens of the country. I am, dear sir, truly yours,

P. H. Sheridan, *Lieutenant General*

St. Paul, February 6, 1879

My dear Sir:— ... it seemed a mystery how campaigns on a large scale, had ever been conducted without their aid [operators]. There can be no doubt that in the late war, thousands of lives and millions of treasure were saved by the field telegraph operators. So far as my observation extended they were as a class brave, energetic and faithful young men who deserve well of their country. Truly yours,   John B. Sanborn

Late Bvt. Major Genl. Vol., Comd. In the Field.

3. The Telegraph and the American Civil War

United States Senate Chamber,
Washington, March 3, 1879

My Dear Sir:—Your kind letter of the twenty-fourth inst. is at hand. I have no hesitation in saying that the "Telegraph Corps" was of infinite service during the late war, and I am free to say that I never knew a body of men who possessed more integrity, industry and efficiency than the operators with whom I was thrown. I wish you great success in your work....

Yours truly,   A.E. Burnside

State of New Jersey, Executive Department,
Trenton, March 8, 1879

My Dear Sir:—Your letter of the twenty-fourth ulto has reached me. I am very glad to learn that a competent person has undertaken the task of doing justice to the services of the Military Telegraph operators during the war, and I share with you your feeling of surprise and regret that the work has not already been done. I do not think that anyone appreciates more highly than I do, the value of those services, and the loyal and invaluable devotion so constantly displayed by the men of whom Caldwell was so excellent an example.[30] If I can at any time be of service to you in carrying out your plan, please let me know.

In great haste.   Very truly yours,

Geo. B. McClellan[31]

Why is it that these generals praised the telegraph service when many of them suffered from its shortcomings? Perhaps it was the passage of time or that they believed that the military telegraph was basically a benefit to them. Another question was, why did these problems persist during the war, why were they tolerated?

The legal scholar and historian Roscoe Pound, writing in the *Journal of the Massachusetts Historical Society*, believed that the answer is partly to be found in the Anglo-Saxon instinct for decentralization, checks and balances, and in general accomplishing your goals without assistance. He cites the experience of the British, whose War Office contained a system of bureaus which functioned inefficiently during the stress of battle during the Crimean War, when the telegraph was first used in a conflict. As an example during the Civil War, he states that the Bureau of Military Telegraphs was not the only one that was conducted as an end in itself rather than toward the ends of effective military operations. During the Knoxville campaign a lieutenant of the ordnance department defied General Amos Burnside, then in command of the

Army of the Ohio, by refusing to fill out a requisition for arms needed by one of Burnside's regiments. As a consequence, the army was deprived of the use of that regiment on the eve of battle. When the general had sought to maintain discipline by arresting the lieutenant, telegrams arrived from the War Department vindicating the independence of the lieutenant (since he was from a different command) and rebuking the general. Another point that Pound makes is that these rulings were possibly made by bureau officials other than the person in charge (in most cases concerning the telegraph it was Secretary of War Stanton) to keep as much power as possible within the bureau. General Schofield has said that bureau officials habitually used the name and authority of the secretary.[32] Perhaps it will never be known how much Stanton really had to do with the messages which Colonel Stager or his subordinates sent out in his name.[33]

On the Confederate side, at the beginning of the war, all the private telegraph lines were intact. The major cities and towns of the South enjoyed the same access to information as to the progress of the war as the North. The main difference was that the South did not attach as much importance to the telegraph as a major aid in warfare as did the North. While the South did nationalize the two southern telegraph companies, the American and the South-western, they did not form a military telegraph organization. On May 10, 1861, these companies were combined into the Southern Telegraph Company and placed under the command of Dr. W.S. Morris.

Morris had much the same problems as did Colonel Stager in the North. As an example, Major General D.H. Hill complained that it took six hours to find his telegraph operator at a time when he wished to reply to a critical order from Lieutenant General James Longstreet having to do with "movements of the Yankees." "I do not know the reason of his absence or negligence," growled Hill, "but such things at this time are very alarming." When Major General Sam Jones reported the telegraph operator at Glade Springs, Virginia, to Morris for drunkenness and asked for a replacement, the delinquent operator provided Morris with an interesting explanation:

> In the first place I did not pretend to Operate that night at all, as I had gotten Mr. Pepper a Sounder[34] who happened to be here to attend to my office for me that night so that I might recruit a little ... from the Quantity of business

3. The Telegraph and the American Civil War

done during the month.... I took a drink that night & not having any supper & loss of sleep it made me tight But Mr. Pepper sent all of Genl Jones Telegrams & got the answers required in about one hour & the General left on the Mail going West after receiving all the information desired & for what cause he could have to report me I don't see.[35]

Southern newspapers complained when there were any problems with the military telegraph, such as operators who closed shop too early in the evening to permit important news to reach them, press dispatches that were inaccurately routed, and telegraph operators who took liberties with the facts. The editor of the Lynchburg *Republican* wrote Morris in May 1863 that:

> I am sorry to have to complain to you of the manner in which your office is conducted here. On Friday last an important dispatch about matters at Fredericksburg was sent to the Virginian & not to us. On Saturday night another came about a cavalry fight at Chancellorsville & it was not sent to us until Sunday morning too late for our paper. In the first case they say that they neglected to make a copy & in the next they can give no explanation at all. These grievances occur frequently or I would not complain. The fact is you have not a competent man in your office. There is not one of them who can write half as good as your little son.[36]

The practical problems of the Southern Telegraph Company were in large part the product of wartime scarcities. Increasing costs of labor and materials, defective equipment, the construction and maintenance of telegraph lines menaced by enemy invasion, the procurement of telegraph operators, sporadic cases of disloyalty and other forms of misconduct on the part of telegraph employees, and the specter of incipient trade unionism among the telegraphers had to be dealt with by Morris at one time or another during the war.[37]

The main scarcity was the materials to construct and repair the system during the fighting. At the beginning of the war there was not a single wire or glass factory in the South, and also a crucial shortage of telegraph wire and battery acids. Morris displayed great ingenuity in supplying these deficiencies, obtaining sulphuric acid from all parts of the Confederacy and Mexico and starting a wire factory at the Tredegar Iron Works in Richmond that produced excellent wire for the telegraph.[38] Bluestone (copper sulphate) for batteries was in particularly great demand, some thirteen thousand pounds being consumed each year by the Southern Telegraph Company.[39] As the war progressed, Morris increasingly came to depend on foreign sources of supply and on blockade runners for necessary material, much of which came from England.

These equipment problems plagued Morris throughout the war and caused him to be a continuous target of criticism from both the southern press and the Confederate military. A Savannah newspaper in June 1863 complained of the "rotten and indifferent post and the careless manner in which the wires are attached over and about one half the lines," and when a rain storm or high wind came along, "the whole establishment" went down and "everything like news" was intercepted.[40] Interruptions in the service were frequently the result of the earthenware insulators used on many of the company lines. One of Major General A.B. Magruder's staff officers early in 1862 attributed the repeated failures in communication to these earthenware insulators on a particular section of the line between Williamsburg and West Point, Virginia. He advised Morris that "unless better ones—either of glass or earthenware—better glazed and of superior quality to those in use be furnished—it becomes a question whether it is any longer incumbent on the government or important to the public to cooperate with your company in keeping up the line."[41] Ultimately the officers of the Southern Telegraph Company came to the same conclusion about their insulators. Morris' Alabama superintendent, J.B. Tree, in a message him in 1864, referring to the condition of the Blue Mountain telegraph line in that state, said, "It is really useless to build a line with Earthenware Insulators, capable as these are of an absolute porousness in a heavy rain—and the worst of it is they stay damp so long."[42]

While the equipment problems that Morris had did not occur in the North, both telegraph systems had difficulties with operators. The main southern problem was in finding competent operators to meet the demands of the new lines of the military telegraph. At the beginning of the war, there were not only an insufficient supply of operators, but many were of northern birth, and moved back when hostilities commenced. Others had enlisted in the Confederate army, so that the only remedy lay in detailing operators from the ranks, which was done. In a letter to the Confederate Adjutant General Samuel Cooper written in early 1863, Morris explained that the need for operators has "reduced our force in our regular offices so much that our men are worked night and day to keep up with the urgent demands of business and however diligent they

3. The Telegraph and the American Civil War

may be they cannot do it with as much dispatch as the importance of it demands."[43] On the same subject, Morris contacted the Confederate secretary of war, James A. Seddon, fourteen months later that "it not infrequently happens that the offices are ordered open from night to night for weeks consecutively and a man who has scarcely left his instrument during the day has to sleep with his ear on the Sounder, or not sleep at all."[44] Morris also had to deal with the Confederate congress on the subject of the amount of manpower needed in each location: "I would respectfully state that at Richmond there are five instruments each requiring an operator; at Wilmington N.C. there are four instruments; at Augusta, Geo. five instruments; at Montgomery, Ala. three instruments; and at Lynchburg, Va. also three instruments; and when the military movements are important, additional operators are necessary to perform the night service."[45]

General Robert E. Lee also became involved in the telegraph operator situation because it came to his attention that the operators were passing the time playing cards. Lee wrote to Morris about this in November of 1862: "I have been informed that the practice of card playing prevails in the telegraph office at Richmond among the operators and assistants. Although I do not suppose that it is carried out to an injurious extent; but merely indulged in as an amusement. Yet I have thought it best to call your attention to the fact, in order that you may guard against the evil that might result from it."[46]

The Confederate military also had the same problem as with operators in the North—profiting from their occupation to influence the stock market. One of these accusations was made against an operator in the telegraph office in Charleston, who apparently had been making money in the stock market through the use of early information of the arrival of blockade-runners into Wilmington, North Carolina.

One of the most serious instances of this type involved William Roche, a division superintendent with the Southern Telegraph Company, headquartered in Montgomery. General Superintendent Tree produced evidence to support four charges against him:

1. That at the time of the fall of Atlanta in September 1864, Roche sent a telegram to a W.A. Gardner in Mobile, stating in substance, "Buy from three to four thousand Gold immediately—answer tonight."

2. That between July 13 and July 30, 1864, Roche was in the city of Mobile "beastly drunk" for the greater part of that time and completely incapable of attending to the important interests over which he held control.

3. That Roche entered into a contract with Gardner for the delivery of telegraph poles for the military line to Blue Mountain. Although the poles were never delivered, Roche collected from Morris, as manager of military telegraph lines, the sum of $8,557.50, which he in turn paid to Gardner.

4. That Roche issued a false certificate of exemption to Gardner, stating that he was in the employ of the Southern Telegraph Company.

Roche was convicted of these charges and dismissed from the company, and turned over to the authorities to recover the money that the government had been defrauded of through the pole contract.[47]

In some cases, the southern military, in order to solve their difficulties with the telegraph, began to take charge of the lines within their commands without consulting Morris. In the spring of 1861 Pierre G.T. Beauregard, the first southern general to appreciate the importance of the telegraph, built up a regularly organized telegraphic system with a full corps of operators, whose salaries were higher than those paid to Morris' employees. And Morris himself was not immune to criticism. On November 17, 1861, Major General John B. Magruder complained to the Adjutant General that:

> Dr. William S. Morris, president of the telegraph company, has failed to send forward the materials and chemicals necessary to keep the telegraph lines in operation which I have constructed with so much labor, and ... I have sent Mr. Conner, one of the telegraph operators, to Richmond for the purpose of getting the necessary chemicals and not having returned, so far as I have learned, I presume Dr. Morris has kept him also. For the want of these articles the line is seldom in operation between Yorktown and Richmond even.... I have refused to send the men from this department which Dr. Morris ... requires, and I hope the War Department will sustain me in it, and cause Dr. Morris to send Mr. Conner back, with the necessary chemicals and wire, forthwith.[48]

The Confederate telegraph system had difficulties with trade unionism that arose when a group of telegraphers, most of them employees of the Southern Telegraph Company, met in Augusta in October 1863

3. The Telegraph and the American Civil War

and formed a union they called the Southern Telegraph Association. The grievances that impelled this action were the long hours of duty and salaries "entirely inadequate to defray the expenses of living in the simplest manner." Faced with this threat, Morris ordered those who were his employees to leave the association under pain of dismissal and subjection to military service at a soldier's pay of seventeen dollars a month. After a deadlock of ten days, the operators began to leave and return to duty, causing the strike to collapse. Thus ended the first and only strike in the Confederacy and with it the union of telegraph employees.[49]

A hazard that the southern operators shared with those in the North was the risks and dangers of being in the line of duty. An operator at Guinea's Station near Fredericksburg, Virginia, reported to Morris that the "enemy appeared in full view of Guineas this morning and continued to advance. When within a few hundred yards of the depot I started on a hand cart with the register and magnet. Did not have time to save the balance ... they tried to cut us off ... but failed."[50] Capture and even death were the lot of other operators. District Superintendent Crowley and the Jonesboro telegraph operator were captured during a union cavalry raid at Jonesboro, Tennessee, in September 1863 and incarcerated in a prison in Knoxville for a month and a half.[51] In Wilmington, North Carolina, the manager of the Southern Telegraph Company's office was a victim of yellow fever; he disregarded the company's instructions to leave at the beginning of an epidemic and remained at his post until he succumbed to the disease in October 1862.[52]

At no time during the war did the Southern Telegraph Company show its abilities more than in its desperate attempt to keep its lines in operation during Sherman's march from Atlanta to Savannah. By the time Sherman left Atlanta in November 1864, Morris' company had almost exhausted its supply of telegraph materials. Nearly all the telegraph wire in the Confederacy was on poles, and much of that was virtually useless owing to rust.[53] As Sherman marched eastward, Charles G. Merriwether, then assistant superintendent at Mobile, travelled to Macon, Georgia, as it was being abandoned by the Confederates, packed up all the surplus telegraphic materials and sent them to Montgomery, Alabama. By November 22, all communication from the West to Augusta and Savannah was cut off. On the following day Superintendent Tree telegraphed Augusta: "Keep your offices open night and day. If you have to fall back, take it coolly and gather up the operators, instruments, and material as you retire. If the enemy diverge from the Central or Georgia Road, establish an office at the end of the break and send your business through by couriers. We will do the same at the Macon end."[54]

This message had to go by way of Columbus, Mississippi; Tallahassee, Florida, and Savannah, the only route left open. The next day the line over which the message was sent was cut south of Macon. In spite of valiant efforts, Tree could accomplish little. Merriwether, after rebuilding the line for nine miles east from Macon, was driven back and forced to remain idle for five days during which what Tree described as "probably the most destructive raid known in this war" when the Union forces destroyed everything south of Macon for a hundred miles.[55]

With the end of the war coming, the stalwarts of the Southern Telegraph Company attempted to keep functioning, and at the same time to keep the company's bottom line in mind:

> We are straining every nerve to reopen a communication with you all. See that a corresponding movement is made from your end....
>
> The President of the Macon & Western Railroad has been urging us to rebuild the Line along the Road—I have simply replied that it requires at this time all our materials to Keep open a communication with Virginia and that, when we can do this it will give me great pleasure to talk with him further on the subject....
>
> From persons lately arrived here [Augusta] it would seem that Sherman is marching in such a manner to the Eastward, as to render it doubtful whether he intends Wilmington or Charleston as his ultimate object. The only policy I can pursue under the present status is to Keep in view the importance of reopening a communication with Virginia ... not only as being important to the best interests of the government, but as being vitally necessary to restore our revenue now crippled by the Extensive destruction of our lines by the Enemy.[56]

One of the great advantages that the Union army had during the Civil War was its superior telegraph system. Even with the problems caused by some of its operators, the North's vast communication network was manned by expert telegraphers as well as cipher specialists. The South, in addition to its continuous material problems, simply did not have this level of experienced personnel. The faithful service of com-

3. The Telegraph and the American Civil War

mercial and railroad telegraphers that the Confederacy possessed was no substitute for the North's well organized military telegraph.[57]

The Civil War, through the use of telegraphy, allowed us to see an effective means of military information processing that, coupled with networking, allowed for the centralization of power and control, which in turn significantly altered the logistics of the war.

4

The Telegraph and Abraham Lincoln

On July 21, 1861, when the Battle of Bull Run was fought in Manassas, Virginia, a telegraph line from Washington, D.C., was connected to the Fairfax Court House in Fairfax, Virginia. From that point, the information was carried by horseback to the headquarters of Brigadier General Irwin McDowell, the commander of the Union forces. Information about the battle was send back to Washington the same way. Initially, the telegrams from McDowell during the morning were encouraging. By the afternoon, however, the telegraphic messages ceased. This occurrence caused no alarm to the assembled officials in the War Department, including Abraham Lincoln; the belief was that the Confederates had been routed and that the telegraph's capability had been outpaced. Unfortunately for the Union, the reverse was true, the next to last message from the field read, "Our army is retreating." And finally, "the day is lost, save Washington and the remnants of this army."[1]

From this point on Lincoln became interested in the telegraph and the information on the fighting that it provided, and this interest became more intense as the war dragged on. Some of this intensity was caused by his subordinates. On October 21, 1861, a message arrived via telegraph to McClellan's headquarters in Washington. The message informed McClellan of the Union loss at the Battle of Balls Bluff. While it was not a huge loss for the Union, it was the second battle success in a row after Bull Run for the Confederates and one of the Federal dead was Colonel Edward D. Baker, a United States senator from Oregon and a friend of Lincoln's.

Because McClellan had little faith in Lincoln's military judgment, he did not inform him of the results of the battle. Later that day, Lincoln visited the War Department's telegraph office and inquired if any dis-

4. The Telegraph and Abraham Lincoln

patches had arrived from the front. Thomas Eckert, in charge of the office, had been ordered to give military messages only to McClellan, so, caught in a bind, Eckert slipped his copy of the dispatch under his blotter and told Lincoln that there were no new dispatches on file. Lincoln, leaving the building, saw a copy of a military dispatch on McClellan's desk giving the results of the battle. From that time on, when he was told that there was no war news, he would sometimes say: Is there not something under the blotter?"[2]

Either because he believed that all relevant information was being kept from him or because the war was not going well, by the second half of 1861, Lincoln turned to the telegraph to project his leadership. In 1862, he began issuing direct orders to his generals in the field, beginning a new phase in warfare—the direction of military leaders by a political leader in real time, a feat only available at that time by the use of the telegraph.

Because there was no precedent, Lincoln, through sheer ability, discovered how to use the telegraph on his own to guide the Union to victory. Previous to the use that Lincoln made of the telegraph, the military used it to order and ship supplies and to move personnel. Part of Lincoln's genius was in his ability to master innovative technological ideas. As a lawyer in Illinois, Lincoln gained experience in dealing with the other great nineteenth century innovation, the railroad.

Lincoln had a genuine intellectual curiosity about new technology. Two years before he was elected president, he gave a series of lectures entitled "Discoveries and Inventions." Its first words were, "All creation is a mine, and every man a miner." Technological innovation and its benefits, Lincoln stated, were what separated "Young America" from "Old Fogey" other nations to the advantage of the new republic.[3]

One of Lincoln's first acts concerning the telegraph was to remove its control from the military and to place it in the hands of the War Department; now communication with the military would be controlled by civilians. The War Department was next to the White House, giving Lincoln easy availability to the technology. The telegraph now became one of the centers of the president's activities. "His tall homely form could be seen crossing the well-shaded lawn between the White House and the War Department day after day with unvaried regularity," wrote one observer.[4]

Other than in the Executive Mansion, Lincoln would spend most of his time here in a small office next to Secretary Stanton's, running the Union war effort. He would "take his pen or pencil in hand, smooth out the sheet of paper carefully and write slowly and deliberately, stopping at times in a thoughtful mood to look out of the window for a moment or two, and then resuming his writing."[5]

The president devised a simple method of dealing with the incoming military information. He would open the telegraph operator's draw and leaf through all the messages that had come through since his last visit. In this way, he would have a complete picture of the progress of the war. If there was breaking news, he would hover over the telegraph operators reading the dispatch word by word as it was decoded.

Lincoln also had a habit of walking to the government departments that were clustered near the White House. In this way, he could have face to face conversations with those who ran the various parts of the administration. The telegraph gave him the same ability to "converse" with his generals in the field. While he did not avoid the chain of command, his messages added any emphasis that he wished to impart depending on the situation.

Early in January 1862, Lincoln began to make his presence felt in the military theater. General McClellan had come down with typhoid fever, leaving his two subordinate generals—Henry Halleck in the West and Don Carlos Buell in Tennessee—leaderless. At this point, Lincoln was still using letters more than the telegraph, especially if the contents were lengthy. To each general he telegraphed that he would write to them. In these slower messages, Lincoln urged an advance against the Confederates, closing the letters with, "Please do not lose time in this matter."[6]

A few days later Lincoln telegraphed Buell inquiring, "Have arms gone forward for East Tennessee? Please tell me the progress and condition of the movement in that direction."[7] Since the general wired back excuses, Lincoln again used written communication to explain the basis for his questions and deferred to the general by closing with "I do not intend this to be an order in any sense."[8]

But by the end of January with no movement from the field, Lincoln's deference evaporated. The message to both commanders read, "Please name as early a day as you safely can, on, or before which, you

4. The Telegraph and Abraham Lincoln

can be ready to move Southward in concert with General Buell [Halleck]. Delay is ruining us; and it is indispensable for me to have something definite. I send a like dispatch to Buell [Halleck]."[9]

The result of these telegraphic dispatches was not a southward move, but the resurrection from his sickbed of General McClellan who made a surprise visit to the White House on January 12, "fearful that his command was being undermined" by Lincoln.[10]

McClellan then put together his grand plan to attack Richmond by moving his army by boat to Virginia just south of the Confederate capital near Yorktown. It was a brilliant plan and well executed. Unfortunately it was conceived by McClellan. The Union force of sixty-seven thousand men faced a Rebel force of only thirteen thousand, but the Confederate commander, General John Magruder, fooled McClellan into believing that his army was much larger, too large in McClellan's mind to attack.

With McClellan immobile, Lincoln moved General McDowell's force of thirty-eight thousand men to Washington to defend the capital instead of sending them to McClellan, as was the initial plan. At the same time, he had been reading McClellan's telegrams to the War Department and feared that "indefinite procrastination" was the general's plan. Frustrated, he asked, "Is anything to be done?"[11]

Lincoln now took the next step in his evolution to become commander-in-chief. He boarded a ship, sailed south, and ordered the shelling of the Rebel positions around Norfolk, Virginia, and then watched Union forces take the city. As a bonus, he caused the destruction of the Confederate ironclad *Merrimac*, which was destroyed by the cities' defenders lest it fall into enemy hands. Salmon Chase, the treasury secretary, said of Lincoln's adventure, "So ended a brilliant week's campaign of the President, for I think it quite certain that if we had not come down, Norfolk would still have been in possession of the enemy, and the Merrimac as grim and defiant and as much a terror as ever."[12]

Back in the White House, Lincoln now retuned to dealing with McClellan, who, with his army which had grown to eighty thousand, still believed that he was opposed by a larger force. Additionally, Confederate General Stonewall Jackson, after defeating the Union force at the Battle of Front Royal, was moving through the Shenandoah Valley towards Washington. To deal with both situations, Lincoln now gave direct commands to his generals.

To deal with the Jackson threat, Lincoln sent the following wire to General John C. Fremont, stationed with a small force in western Virginia: "The exposed position of General Banks make his immediate relief a point of paramount importance. You are therefore directed by the President to move against Jackson at Harrisburg.... This movement must be made immediately." Instead of any pleasantries, the telegram ended with, "You will acknowledge the receipt of this order and specify the hour it is received by you."[13]

Also in Virginia, at Fredericksburg, General McDowell received a message from Lincoln, "Gen Fremont has been ordered by Telegraph to move from Franklin on Harrisonberg to relieve Gen Banks and capture or destroy Jackson & Ewell's force. You are instructed ... to put twenty thousand men (20000) in motion at once for the Shenandoah.... Your object will be to capture the forces of Jackson & Ewell."[14] McDowell telegraphed back, "I have ordered General Shields to commence the movement by tomorrow morning."[15] The president replied, "Everything now dependes [sic] upon the celerity and vigor of your movement."[16]

To McClellan, Lincoln telegraphed, "In consequence of Gen Banks critical position I have been compelled to suspend Gen McDowell's movement to join you. The enemy are making a desperate push upon Harper's Ferry, and we are trying to throw Fremont's force & part of McDowell's in their rear."[17]

On paper Lincoln's moves, if successful, would trap Jackson, but Jackson moved too quickly and on May 25, he hit Banks at Winchester, Virginia, defeating him so badly that Banks had to leave his supplies behind. Now Lincoln moved back to McClellan, becoming very direct in his instructions. "I think the time is near when you must either attack Richmond or give up the job and come to the defense of Washington. Let me hear from you instantly."[18] McClellan did reply quickly that he was looking to attack Richmond, on his own schedule.

The president, however, had to keep his focus on Jackson, still believing that he could trap him between Fremont and McDowell. But it was not to be since Fremont was not on schedule; he was in fact twice as far from Harrisonburg as he should have been, a necessary detour according to Fremont because of bad weather, bad roads, and because he had to feed his men. With his plans to trap Jackson temporarily awry,

4. The Telegraph and Abraham Lincoln

the President wired McClellan promising support, but at the same time establishing that he, Lincoln, was in charge.

"That the whole force of the enemy is concentrating in Richmond, I think can not be certainly known to you or me. Saxton at Harper's Ferry, informs us that a large force (supposed to be Jackson's and Ewell's) forced his advance from Charlestown today. Gen. King telegraphs us from Frederick'sburg that contrabands give certain information that fifteen thousand left Hanover Junction Monday morning to re-inforce Jackson. I am painfully impressed with the importance of the struggle before you; and I shall aid you all I can consistently with my view of due regard to all points." And the closing, "And last I must be the Judge as to the *duty*, of the government in this respect."[19]

Lincoln still believed that his three generals—McDowell, Fremont, and Banks—could stop Jackson, but it was not to be. Jackson not only slipped out of the trap by pushing his men harder than the Federals did, but turned on his pursuers, defeating Fremont at the Battle of Cross Keys and General James Shields' force, formerly McDowell's, the next day at the Battle of Front Royal on his way to aid in the defense of Richmond.

McClellan, still in front of Richmond, was still complaining about his inferiority in numbers, but the battle for the Confederate capital was now underway and McClellan was winning, and against Robert E. Lee. This was the Battle of the Seven Days; Lee would attack, McClellan would stop him and then McClellan would retreat. Worse, McClellan blamed Lincoln for the situation: "If I save this Army now I tell you plainly that I owe no thanks to you or any other persons in Washington—you have done your best to sacrifice this Army."[20] Secretary Stanton and President Lincoln never saw the last sentence. The supervisor of the telegraph office was so taken aback by its insubordination that he deleted it from the copies that went to the secretary and the president.[21]

By July 2, the Battle of the Seven Days was over. The field belonged to Lee and McClellan put his troops back on the ships and returned North. For Lincoln, he found that he could command armies using the new invention, the telegraph, but the execution of the commands depended upon those commanded and in this he had a huge problem finding generals who would fight and fight successfully.

With the operations against Richmond concluded without success,

How the Telegraph Changed the World

Lincoln now used the telegraph not for giving military orders, but for gaining information. The president now realized that he was fighting a two-front war. After the second Battle of Bull Run, Lee turned North into the state of Maryland shadowed by McClellan, and Confederate General Braxton Bragg was moving from Tennessee into Kentucky, possibly to join Lee. But was he? Lincoln now used the telegraph to try to find Bragg. At this point, September 17, McClellan finally engaged Lee at the Battle of Antietam, and while the battle, although bloody, was not decisive—Lee escaped, which is the reason Lincoln sent no congratulatory reply to McClellan's telegram to the War Department: "I have the honor to report that Maryland is entirely freed from the presence of the enemy, who has been driven across the Potomac."[22]

On October 8, after several weeks with no telegrams, the president finally sent one of congratulations, not to McClellan, but to General Ulysses S. Grant, who had defeated the rebels at Corinth, Mississippi. As far as McClellan was concerned, the president believed that his period of inactivity had become permanent. After the midterm elections, Lincoln replaced McClellan with General Ambrose Burnside.

The substitution brought no relief to the Union command problems. McClellan had been fired because of his failure to attack; Burnside attacked and his troops were slaughtered at the Battle of Fredericksburg. Once again the Federals were defeated by Lee. On December 30, the president sent a telegram to his new commander that was opposed to his usual prompt to engage the enemy: "I have good reason for saying you must not make a general movement of the army without letting me know."[23]

Since Burnside was not the right commander for the Union army, Lincoln replaced him with General Joseph Hooker in April of 1863. Hooker had distinguished himself fighting with McClellan and initially seemed to be Lincoln's type of general. He immediately began to rebuild the dispirited Union army and with Lincoln's approval set up a plan to move forward. When he put the plan in motion, which if successful would have trapped Lee, Lincoln tried to use the telegraph to stay informed, but received little back from Hooker, who was busy being defeated by Lee in what became the Battle of Chancellorsville.

Even though Hooker had lost at Chancellorsville by not being aggressive enough against Lee, Lincoln decided to stay with him. After

4. The Telegraph and Abraham Lincoln

the battle, with Lee moving northward through May into June, Lincoln again used the telegraph to gain information which he communicated with Hooker. On June 15 Lincoln wired Hooker that Lee had crossed the Potomac and that "I would like to hear from you." Hooker responded that evening, "Your telegram of 8:30 received, it seems to disclose the intentions of the enemy to make an invasion, and, if so, it is not in my power to prevent it."[24]

At this point Lincoln had to believe that Hooker was another McClellan—he controlled the entire Army of the Potomac and he would not use it. Meanwhile Lee continued to march northward. On June 27, Hooker asked to be relieved from his position and Lincoln rapidly complied.

Hooker's successor was General George Gordon Meade, a corps commander under Hooker, and three days later he clashed with Lee at the Battle of Gettysburg. The result was the following telegram from the War Department on July 4, 1863:

> Washington City, July 4, 10 a.m. 1863
>
> The President announces to the country that news from the Army of the Potomac, up to 10 p.m. of the 3rd, is such as to cover that Army with the highest honor, to promise a great success to the cause of the Union, and to claim the condolence of all the many gallant fallen. And that for this, he especially desires that on this day, He whose will, not ours, should ever be done, be everywhere remembered and reverenced with professional gratitude.[25]

Now, with this great victory, was the time to crush Lee. Tremendous rains had blocked his crossing of the Potomac and the safety of moving south. But it was not to be. Meade moved too late to trap Lee, who escaped back to Virginia. General Halleck wired the president's displeasure to Meade: "The escape of Lee's army without another battle has created great dissatisfaction in the mind of the President, and it will require an active and energetic pursuit on your part to remove the impression that it has not been sufficiently active heretofore."[26]

Lincoln's disappointment with Meade was tempered by Grant's success in taking Vicksburg, a seemingly impregnable rebel stronghold on the Mississippi. But ten weeks later, the Union army was again in trouble. General William Rosecrans was defeated at the Battle of Chickamauga. Even though he had lost the battle, Lincoln sent him an encouraging wire on August 31: "I repeat that my appreciation of you has not abated.

I can never forget, whilst I remember anything, that about the end of last year and beginning of this, you gave us a hard earned victory, which had there been a defeat instead, the nation could scarcely lived over. Neither can I forget the check you so opportunely gave to a dangerous sentiment which was spreading in the North."[27]

By these words, Lincoln was repaying an important debt. Eight months earlier Rosecrans had won an important victory at the Battle of Stone River. At that point the belief in the administration was that if the South had one more substantial victory before the end of 1862, nations such as England and France would come into the war on the Confederate side. Rosecrans' victory prevented that eventuality.

The president was on the telegraph day and night gaining information and giving orders; the victories at Vicksburg and Gettysburg could be offset if the South gained control of the middle of the country. The action now centered on the city of Chattanooga, to where the Union forces had retreated. Once again Lincoln would be let down by his bickering incompetent generals. In this case it would be Burnside, who avoided Lincoln's orders, and Rosecrans, who did not wish to have his force intermingled with those of Hooker, who was coming to his aid. This aid proved the value of the telegraph, used for coordination, and another early nineteenth century invention, the railroad. The Union army amassed two army corps totaling 23,000 and all their supplies, put them on a train at Bealeton, Virginia (thirty-seven miles south of Washington), and sent them 1233 miles to Chattanooga, Tennessee, in eleven and a half days. Because of their agents in Washington, the Confederate army knew of this action, but could not stop it.

Orange Court House, Va., Oct. 3, 1863
Hon. Jefferson Davis, President, Richmond, Va.

A dispatch from Major Gilmor last night states that reinforcements for Rosecrans have all passed over the Baltimore & Ohio Railroad. The force composed of Slocun's and Hooker's corps, estimated at between 20 and 25,000 men. He states he made several attempts to break the railroad but could accomplish nothing....

R.E. Lee, General.[28]

With this force in place, the president again made a change at the top of the command structure. He created the Military Division of the Mississippi and placed Ulysses Grant in charge.

4. The Telegraph and Abraham Lincoln

Grant immediately made changes. He relieved Rosecrans, replacing him with General George Thomas, and placed General William Tecumseh Sherman at the head of the Army of the Tennessee. Together with Hooker, the new force drove the Confederates from Chattanooga and opened the way to the deep South.

The following day Lincoln telegraphed Grant, "I wish to tender you, and all under your command, my more than thanks—my profoundest gratitude—for the skill, courage, and perseverance, with which you and they, over so great difficulties, have affected that important object. God bless you all."[29]

The president finally had both an able commander and a means of not only looking at the actions in the field (Lincoln saw all of the telegraphic traffic between Grant and his generals), but controlling them.

A field telegraph station at Wilcox's Landing, Virginia, in the vicinity of Charles City Court House, 1864 (Library of Congress).

How the Telegraph Changed the World

By 1864, the uses of the telegraph in war expanded to include not only the fighting, but also the management of ancillary operations such as the manufacture of material and the planning of railroad movements. The telegraph also allowed military commanders to remain in the field and moving with their armies instead of staying in one place. As Grant wrote to his father, "In these days of telegraph & steam I can command whilst traveling and visiting about."[30]

Grant commanded his forces as Lincoln commanded Grant. In each case, the main objective was agreed upon and the execution left to the subordinate. Lincoln told Grant the war needed to be kept in the South, Lee had to be kept below the Potomac and finally that the Confederates had to be continuously engaged and defeated. From Grant to Sherman in the West, "You I propose to move against Johnson's army, to break it up and to get into the interior of the enemy's country as far as you can.... I do not propose to lay down for you a plan of campaign, but simply to lay down the work it is desirable to have done, and leave you free to execute it in your own way." From Grant to Meade, "Lee's army will be your objective point. Wherever Lee goes, there you will go also."[31]

On May 4, 1864, Grant and Meade crossed the Potomac and marched South. It was an open secret which Lincoln himself shared that Grant preferred to be cut off from Washington while on this march. A Lincoln wire to Grant on April 30, 1864 stated: "The particulars of your plans I neither know nor need to know."[32] Therefore for nearly a week there was no telegraph communication until the Battle of the Wilderness, a forest area that nullified the Union advantage in numbers and artillery. While the Federal forces lost the battle, the important fact was that afterwards they did not retreat, but rather marched South pursuing Lee. Lee continuously fell back, entrenched his forces and engaged the Union, winning every battle, but losing irreplaceable manpower and supplies. Lee then did what Lincoln feared most, he moved North, sending General Jubal Early and fourteen thousand troops towards Washington.

Until this point, Lincoln's telegraphic messages had been in the nature of monitoring Grant's movements, but now he needed Grant and his military force to protect the capital. However, he did not order Grant to come to Washington personally, but merely suggested. Grant missed the "suggestion" about accompanying his troops, but wanted to solve Lincoln's problem. His reply, "I have sent from here a whole corps com-

4. The Telegraph and Abraham Lincoln

manded by an excellent officer, besides over three thousand other troops.... They will probably reach Washington tomorrow night."[33] Because General Lew Wallace had held up Early at the Monocacy River, forty miles outside Washington, Grant's corps had time to come to his aid. Now outnumbered, Early retreated. Never before during the war had a Confederate army been so close to Washington. This could have been a major disaster for the nation. In an address in May 1902, Leslie M. Shaw (then secretary of the treasury) stated, "With the national capital in the hands of the enemy it would have been impossible to prophesy the foreign complications, to say nothing of the demoralization of the people of the United States." Grant has said of this raid, "If Early had been one day earlier he might have entered the capital."[34]

The next situation involving Lincoln, Grant, and the telegraph was political rather than military. Jubal Early had moved South almost taking Washington, but then turned back North burning down Chambersburg, Pennsylvania. Lincoln believed that the rebels were again setting up to attack Washington and for political reasons, that a general needed to be in charge of the Union military forces around the capital to make tactical decisions, something that Grant could not do from afar. This situation comes back to Lincoln's use of the word "suggestion" in his previous telegram to Grant to have him take charge of the capital troops. This time the president and the general-in-chief met face-to-face. Since Grant wished to continue to pursue Lee, Lincoln would be forced to put Halleck in charge of Washington. Grant did not want this: "I want Sheridan put in command of all the troops in the field, with instructions to put himself South of the enemy, and follow him to the death. Wherever the enemy goes, let our troops also go." Lincoln agreed, but admonished Grant, "I repeat to you it will neither be done nor attempted unless you watch it every day, and hour, and force it."[35]

Sheridan pursued Early through the Shenandoah Valley, defeating him in battle after battle. Lincoln, however, was concerned because Early had escaped before and turned North. This part of the war finally came to an end when a reinforced Early attacked Sheridan at Cedar Creek and was completely routed.

With this victory Lincoln could now congratulate the winning general without fear that the problem would reoccur. "With great pleasure I tender you and your brave army, the thanks of the Nation, and my own personal admiration and gratitude, for the month's operations in the Shenandoah Valley, and especially for the splendid work of October 19, 1864."[36]

Lincoln could also support a general, even when it looked as if the general was not performing, even going against Grant' s wishes. This situation concerned General Thomas, who was defending Nashville from Confederate attack by General John Bell Hood. Since Thomas had the larger force, Washington believed that he should attack Hood. Lincoln's initial belief was that "this seems like the McClellan and Rosecrans strategy of do nothing and let the rebels raid the country."[37] On December 9, Grant sent a telegram to Halleck to order General J.M. Schofield to replace Thomas. The order, however, was suspended by Lincoln. On December 13, Grant wrote a second letter relieving Thomas; this message was to be delivered by hand. Finally on December 15, the weather (which had prevented Thomas from attacking) cleared up, allowing Thomas to engage Hood and won one of the largest victories achieved by the Union during the Civil War.

Lincoln also used the telegraph to involve the military in his reelection campaign. To Grant, a month before the election, "Pennsylvania very close, and still in doubt on home vote—Ohio largely for us ... Indiana largely for us.... Send what you may know of your army vote."[38] Some states would only allow soldiers to vote if they returned to their home state. On September 19, Lincoln telegraphed Sherman: "The State election of Indiana occurs on the 11th of October, and the loss of it to the friends of the government would go far towards losing the whole Union cause. The bad effect upon the November election, and especially the giving the state government to those who will oppose the war in every possible way, are too much to risk.... Indiana is the only important State, voting in October, whose soldiers cannot vote in the field. Any thing you can safely do to let her soldiers, or any part of them, go home and vote at the State election, will be greatly in point."[39]

Lincoln's electioneering efforts with the military were unnecessary as it turned out; his victory margin was overwhelming. Additionally, the president had carried a majority of the military votes against their old commander, George McClellan.

Lincoln's victory also in effect told the South that there would be no negotiated peace which would allow the South to become an inde-

4. The Telegraph and Abraham Lincoln

pendent nation. On January 31, 1865, Grant forwarded to Lincoln the text of a message he received from three high ranking representatives of the Confederate government who wished "to proceed to Washington to hold a conference with President Lincoln upon the subject of the existing war, and with a view of ascertaining upon what terms it may be terminated."[40] For the next two months, the president was involved in ongoing talks with the rebels, but these talks did nothing to deter Lincoln from pressing ahead militarily. On February 4, Secretary of War Stanton wired General Grant, "The President desires me to repeat that nothing transpired, or transpiring with the three gentlemen from Richmond is to cause any change, hindrance or delay, of your Military plans or operations."[41]

After his reelection, Lincoln travelled to City Point, Virginia, Grant's headquarters near the town of Petersburg. Here, the president spent his time reading the telegrams from Grant who was with his force in the field. Lincoln also became the link between Grant and the War Department, relaying military dispatches to Stanton. The Union forces were moving swiftly. On April 3, they took Petersburg, and on April 4, the president was sitting at Jefferson Davis' desk in the newly captured city of Richmond. The day before, the president created a very simple cipher telegram, but so very transparent that it hardly deserved to be called a cipher. The message was sent as follows:

Headquarters Armies of the U.S., City Point,
8:30 a.m., April 3, 1865
To Charles A. Tinker, War Dept., Washington, D.C.:—

Lincoln its in fume a in hymn to start I army treating there possible if of cut too forward pushing is he is so all Richmond aunt confide is Andy evacuated Petersburg reports Grant morning this Washington Sec'y War.

[Signed] S.H. Beckwith[42]

On April 6, Sheridan, closely pursing Lee, reported that "if the thing is pressed I think that Lee will surrender. Lincoln then wired Grant, "Gen. Sheridan says 'If the thing is pressed, I think that Lee will surrender.' Let the *thing* be pressed."[43] Lincoln left for Washington on April 8. before he arrived at the capital Lee surrendered.

[Special Field Orders, No. 54]
Headquarters Military Division of the Mississippi in the Field
Smithfield, N.C., April 12, 1865

How the Telegraph Changed the World

The General commanding announces to the army that he has official notice from General Grant that General Lee surrendered to him his entire army, on the 9th inst., at Appomattox Court House, Virginia.

Glory to God and our country, and all honor to our comrades in arms, to whom we are marching!

A little more labor, a little more toil on our part, the great race is won, and our government stands regenerated, after four long years of war.

W.T. Sherman,
Major-General Commanding[44]

The North began to win the Civil War when Abraham Lincoln picked the right man to lead his forces, supported by the overwhelming material advantage possessed by the Union. One of those advantages was the telegraph and Lincoln's ability to know when and how to use it and when not to use it. Looking at the war as a commander-in-chief, as no political leader before him did, by reading dispatches from all fronts, allowed him a complete picture of events so that he could make his military decisions. At times, possibly too heavily, "some well-meaning newspapers advise the President to keep his fingers out of the military pie."[45] Lincoln also looked for information: "Do Richmond papers of 6th say nothing of Vicksburg? Or, if anything, what?"[46] He gave orders, this one to Fremont to attack Stonewall Jackson: "Much—perhaps all—depends upon the celerity with which you can execute it [the order]. Put the utmost speed into it. Do not lose a minute."[47] And sent congratulations, here to Grant on taking Petersburg: "Allow me to tender to you, and all with you, the nation's grateful thanks for this additional, and magnificent success."[48] The president also knew when the telegram was not the way to deliver his thoughts, as when he visited his generals in the field. There is no doubt, however, that the telegraph allowed Lincoln to use a new technology to enhance the leadership that won the Civil War.

5

The Atlantic Cable

In 1843, Samuel Morse wrote to Secretary of the Treasury John C. Spencer declaring: "The practical inference from this law [of telegraphy] is that a telegraphic communication on the electro-magnetic plan, may with certainty be established across the Atlantic Ocean. Startling as this may seem now, I am confident the time will come when this project will be realized."[1]

Ten years after this prediction, an Englishman by the name of Frederick Newton Gisborne had a chance meeting at the Astor Hotel in New York with Matthew Field, a civil engineer and the brother of a New York paper manufacturer Cyrus Field. Gisborne was a founder of one of Canada's first telegraph companies. Realizing that New York was 1,200 miles further from Europe than St. John's, Newfoundland, he formed the Nova Scotia Telegraph Company to link Newfoundland to Nova Scotia by a combined overland and undersea cable. The idea was to capture European news from vessels passing on the way south to New York and then telegraph the news south to New York before the boat could arrive.[2]

Gisborne failed, defeated by the harsh terrain, inadequate cable technology and lack of funds, and now in New York, looking for backers for his vision, through Matthew Field, he meets Cyrus Field.

Gisborne took Field through his misadventures in Canada, and while Field could see the efficacy of connecting Newfoundland with Nova Scotia, he began to look past this venture, realizing that a telegraph line from St. John's to Manhattan would make more sense if it could be coupled with a steamer shuttle service between Europe and St. John's. With telegraphic rights, the company that set up this plan would make a fortune. The next mental leap took Field from a Newfoundland–New York line to one running from Europe to Newfoundland and then to New York.

To see if the idea was at all practical, Field contacted two men. One was Samuel Morse, who simply repeated what he had said ten years earlier. The other was an American Navy Lieutenant, Matthew Fontaine Maury, who was superintendent of the U.S. Naval Observatory and who agreed with Morse. Maury, using funds appropriated by Congress, began experiments to find out the depth of the ocean between Europe and North America, the composition of the Atlantic seabed, and whether or not the depth was constant. On February 22, 1854, he sent his conclusions to Secretary of the Navy James C. Dobbin:

> From Newfoundland to Ireland, the distance between the nearest points is about 1600 [nautical] miles and the bottom of the sea between the two places is a plateau which seems to have been placed there especially for the purpose of holding the wires of a submarine telegraph, and of keeping them out of harm's way. It is neither too deep nor too shallow, yet it is so deep that the wires once landed will remain forever beyond the reach of vessels' anchors, icebergs and drifts of any kind, and so shallow that the wires may be readily lodged upon the bottom.
>
> The depth of the plateau is quite regular, gradually increasing from the shores of Newfoundland to the depth of 1500 to 2000 fathoms as you approach the other side.[3]

Maury also sent a copy of his report to both Field and to Morse. Morse had already agreed to meet with Field to discuss an Atlantic cable and the report simply made the project more feasible. Both men had what the other lacked. Field had the money and the contacts to raise more, Morse the technical know-how. Additionally, Morse, like Eli Whitney, had not prospered financially from his invention.

Field now searched for fellow investors. His bona fides included Morse's background in telegraphy and Maury's finding. His first partner was Peter Cooper, who had made a fortune as an inventor and industrialist. He was one of the wealthiest men in America. In addition to establishing his technical college, Cooper Union, he had designed and built America's first locomotive, the *Tom Thumb,* and added to his wealth with interests in iron, coal, and glue, as well as establishing the Baltimore and Ohio railroad. His reaction to Field's proposal:

> It was an enterprise that struck me forcibly the moment [Field] mentioned it. I thought I saw in it ... a means by which we could communicate between two continents, and send knowledge broadcast all over the world. It seemed to strike me as though it were the consummation of that great prophecy, that "knowledge shall cover the earth, as the waters shall cover the deep,"

5. The Atlantic Cable

and with that feeling I joined him ... in what then appeared to most men a wild and visionary scheme; a scheme that fitted those who engage in it for an asylum where they could be taken care of as little short of lunatics.[4]

The next partner was Moses Taylor, another wealthy New Yorker, who made his money as an importer and later became the president of City Bank. Taylor agreed to participate if Field could induce others to join. To this end, he introduced Field to Marshall Owen Roberts, who owned the U.S. Mail Steamship Company. Roberts liked the idea immediately and joined the consortium. The next prospect was Chandler White, who like Field made his money in the papermaking industry. White had unsuccessfully invested with Field in the past, but Field still persuaded him to join.

With the group of investors now complete, they called in Gisborne to examine what had been done before, what needed to be done, how much it would cost, and the expected return on their investment.[5] Satisfied with the answers, these five men agreed to form the New York, Newfoundland, and London Telegraph Company. Part of the arrangement was to buy Gisborne's company at a cost of $40,000 and to also assume $50,000 of Gisborne's debts. The group also made deals with the political leaders of Newfoundland and Labrador to erect telegraph cables on their territories.

The company plan had two phases, the first to connect Newfoundland to New York by cable, the second to connect Newfoundland to Ireland, also by cable.

At this point, Field realized that in order for the venture to succeed, the company would need a great deal of help. For this, he first turned to the Navy, again in the person of Lieutenant Maury, whom he asked for two vessels to lay the cable; however, with the commencement of the Crimean War, no assistance in any form was forthcoming from the Navy. The next assistance came strictly from hard won experience. Field journeyed to England to purchase cable to be used in the completion of the North America line between Newfoundland and New York in the areas where an underwater cable had to be used. He was fortunate to be introduced to John Brett, the head of the Magnetic Telegraph Company, who understood undersea cable technology and the best manufacturers of the cable itself. Field purchased from a London manufacturer eighty-five miles of cable, which was shipped to Canada.

At this point Field's lack of cable knowledge came into play, causing blunders in attempting to lay the line. The mistakes, mainly due to poor seamanship on the part of hired ship captains, caused the loss of the cable and $350,000. Field now returned to England and convinced the cable company to lay the cable in Canada at their own risk, using their equipment and know-how. In this, they were successful; the underwater cable was connected to the landlines and now telegraphic messages could be sent a distance of over one thousand miles from St. John's, Newfoundland, to New York City. Later Peter Cooper would say, "It was in great measure due to the indomitable courage and zeal of Mr. Field inspiring us that we went on and on until we got another cable across the gulf."[6] Finishing the Atlantic side of the venture was not only a major accomplishment. It cost the investors more than a million dollars, but now they had a source of income.

With the North American side of the venture completed, Field returned to London to make preparations for the Atlantic Ocean cable. Because of the mistakes made in the past, Field approached this part of the enterprise cautiously. It was now 1856 and three years had passed since the last ocean soundings had been made. New soundings were made, and the findings made with a new instrument, the Massey indicator, differed from the original report. And a new review of the entire situation ordered by U.S. president James Buchanan supported Maury's conclusions. The British also agreed with his conclusions after doing their own soundings of the Atlantic between Ireland and Newfoundland in 1857: "This space has been named by Maury the telegraphic plateau, and although by multiplying the soundings on it, we have depths ranging from fourteen hundred and fifty to twenty four hundred fathoms, these are comparatively small inequities in its surface, and present no new difficulty to the project of laying the cable across the ocean."[7]

The company now had solid answers to the situation under the Atlantic. What they now lacked was money and Field again traveled to England, this time looking for more investors.

On September 26, 1856, Field, in order to begin looking for investors, first formed another concern, the Atlantic Telegraph Company, and secondly had to prove to British scientists that his idea was feasible. To accomplish the latter, he changed the wire inside the cables to one that was stronger, but would still carry signals. Also the induction coils and

5. The Atlantic Cable

receiving magnets were improved over earlier models. Field also saw the *Great Eastern* for the first time—it was still being built—and realized that only a ship of this size could possibly lay his cable across the Atlantic.

In October, the company issued 350 shares of stock at a price of 1,000 pounds each. By December, all the shares were sold. Field's next stop was the British government. Instead of investing, however, Her Majesty's government decided that it wished to be a customer. Its offer was to furnish ships of the Royal Navy to "take what soundings may be considered needful" and to "favorably consider any request that may be made to furnish aid by their vessels in laying down the cable."

The government also agreed to pay the company 14,000 pounds a year (4 percent on capital) as remuneration for government messages until such time as the net profits of the company attained 6 percent, after which the annual payment would be reduced to 10,000 pounds per annum for the next twenty-five years "from the time of the completion of the line, and so long as [the cable] shall continue in working order." In the blink of an eye, the Atlantic Telegraph Company secured its first customer—provided the cable worked.[8]

One of the experts that Field consulted in England was Samuel Morse, who realized the value of his name, influence, and patents to the enterprise. And if the project succeeded, he would gain a vast sum from the use of these patents. As this was the case, he urged that the cable be laid as quickly as possible because he needed money.

At this point Field took a gamble in purchasing the cable. There was a great deal of discussion and disagreement among Field's advisors as to the composition of the undersea cable. The one chosen cost 225,000 pounds of the 350,000 pounds that had been raised.[9]

With the cable chosen, Field returned to the United States, where he began the attempt to have the American government support the Atlantic Telegraph Company as the British did. One of his supporters was Senator William Henry Seward of New York, who reasoned that the Atlantic cable would diminish the chances of war between the two countries. The basic problem in obtaining Congressional support for a bill allowing the United States to become a customer of the company as the British did came from the southern members of Congress, who wanted nothing to do with England. Additionally, southern senators condemned the legislation as unconstitutional. In the end the bill passed by an extremely small margin (e.g., one vote in the Senate). It provided for a minimum subsidy of $70,000 per year until the company's profits equaled 6 percent of the investment; thereafter the annual subsidy would fall to $50,000. The term of the contract: twenty-five years.[10]

An interesting point happening at this time was the relationship between Field and Samuel Morse. Morse had given Field free patent rights and at his own expense, also, Morse had gone abroad to perform valuable experiments on Field's cable—experiments that continued at his own expense. In return Field appointed Morse an honorary director of his company and offered to sell him one or more shares of stock, at par.

Morse thanked Field for his "kind offer" but explained that he had no surplus funds to invest in shares. Additionally, he inquired as delicately as possible of Field "should there not be something." Morse obviously believed that he should be given some compensation for his troubles, especially because friends would ask him how much he stood to gain from the enterprise: "I have been somewhat embarrassed in replying that 'as yet no interest has been definitely assigned, but I had the promise made verbally to me of my friend Mr. Field that when the company was organized, I should proportionately share with the rest. I am in the hands of friends.' This answer has not always satisfied them; they have remarked, 'This is not a business way of doing things.' My reply has been, 'Mr. Field's word is as good as his note.'"[11]

What ever happened, Morse wished to remain part of the Atlantic cable attempt. As such he offered to put aside for a while any request for a financial stake in Field's new British company. "Whatever claims equitable or otherwise I or my friends may think are just on my part upon the company, let them for the present be waived. I shall not thrust before the company, at this moment when the harmonious action of all is necessary to carry forward the enterprise to a successful result, any private or mere personal object to embarrass our united action."[12]

During the spring of 1857, the ships to be used to lay the cable left their ports. The United States sent the *Niagara*, the largest steam frigate in the world, and a second ship, the *Susquehanna*; both ships were extremely impressive. The British sent the *Agamemnon*, their flagship during the Crimean War, the *Leopard*, and a sounding vessel, the *Cyclops*.

5. The Atlantic Cable

After testing the machinery to pay out the cables, all the ships met at the port of Valentia on the West coast of Ireland. And on August 6, they set sail for Newfoundland. The original plan was for the *Niagara* and the *Agamemnon* to depart from Ireland together each bearing half the cable. They would proceed to the mid–Atlantic, where the halves would be joined. One ship would head back to Ireland and the other to Canada. Under the new plan, which was adopted, the ships would leave Ireland together with the *Niagara* laying cable immediately; at mid-ocean the cable end would be spliced to *Agamemnon*, which would lay its cable on the journey to Canada. For two days the convoy had smooth sailing, but on August 9, the weather worsened, causing the strain on the cable to increase, and in addition there was human error in handling the brake on the amount of line released. As a result, the cable snapped, causing 380 nautical miles of wire valued at $500,000 to be lost.

When the shock of the disaster wore off, Field wrote: "The successful laying down of the cable is put off for a short time, but its final triumph has been fully proved by the experience that we have had since we left Valentia. My confidence was never so strong as at the present time, and I feel sure ... we shall connect Europe and America with the electric cord."[13]

Field had no choice but to be optimistic. He had invested $650,000 in this venture, but now he had to raise more money—to pay for an additional 600 miles of cable and to improve the machine used to pay out the wire.

Raising money after the loss of the cable became a huge problem; the British directors did not have additional moneys to invest and the English public had lost faith in the venture. In America, the financial situation was worse—the problems caused by President Andrew Jackson's earlier attack on the central bank now resulted in the Panic of 1857.

Also lost with the cable was public confidence in Field's enterprise in the United States. News of the ruinous accident reached America two weeks after it occurred, evoking both sympathy and skepticism. Many judged the expedition a noble failure: "When we consider the courage which could undertake this Herculean feat," the New York *Tribune* editorialized, "we are almost as proud of our age as if everything had gone on smoothly, and the lightning were now leaping from continent to continent."[14] Others found a sobering lesson for Americans: "We cannot but fear that the success so much hoped for, will not be so easily and so readily attained as our always sanguine people seem to expect."[15]

Another situation concerned Samuel Morse. Treated badly during the first attempt, his position became worse during the second. Once again, Field turned to Morse for help, this time writing a letter to Washington appealing to the government to allow the expedition to use the *Niagara* as they had before to lay the cable.

But before Morse could comply, he suffered another slight. Field's British partners in the Atlantic Telegraph Company declined to appoint him an "honorary director" of the second cable attempt. In doing so the company deprived him of more than another mark of distinction. It was precisely his standing as an honorary director that had entitled him to join the official party aboard the *Niagara*. The decision banished him from the expedition.

Field explained to Morse that under British law only stockholders in a company qualified for directorships, but Morse did not believe him, as he was an honorary director on the first attempt. Morse wrote to his brother, "If they really desired me to be present, as I was last year, they could have easily have found the means of making me an Honorary Director without violating any spirit of the rule."[16]

Fortunately, the needed revenue came from the original British stockholders of the Atlantic Telegraph Company. This time with the new cable, instead of having the ships traveling together from Europe to America, the decision was made to have the two main vessels—the *Agamemnon* and the *Niagara*—meet at a point in mid-ocean, spice the cables that they both carried together and then have each ship return to their respective continent laying wire as they sailed. On June 12 the convoy left Ireland for a voyage that should have taken four or five days. Instead, because of enormous storms that the ships encountered, the trip took sixteen days to arrive at mid-ocean, with damage to all the vessels.

After several attempts to begin their separate journeys because of a broken cable, on June 28, with the cable on both ships spliced together, the ships began the return voyages laying cable as they sailed. By June 30, each ship was approximately 111 miles from the rendezvous point when bad luck intervened—the cable snapped on the *Agamemnon*. As a result, both ships return to Ireland. The company board of directors met to

5. The Atlantic Cable

discuss the situation and because only 300 miles of cable was lost, they decision was made to make one more attempt, but if it failed, there would not be another.

On July 17, the ships again left Ireland bound for the same rendezvous point and on July 29 spliced the cables together to begin their separate voyages. Field's usual buoyancy was beginning to waver: "When I thought of all that we had passed through, of the hopes thus far disappointed, of the friends saddened by our reverses, of the few that remained to sustain us, I felt a load at my heart almost too heavy to bear, through my confidence was firm and my determination fixed."[17]

Each ship carried instruments which transmitted and received signals throughout the entire length of the cable. Since the electronic impulses tested the continuity and insulation of the wire, it also allowed communication between the ships. While there were continuous problems on each ship, such as dead spots in the cable and unexplained communication breakdowns, the voyages continued. The *Niagara* made good headway and would have been even closer to its destination but for being off course, due, it was believed, to the effects of metal cable on the ship's compass. If the compass error had not been discovered, the *Niagara* would have exhausted its supply of cable before reaching land. The *Agamemnon*'s problem was fuel; it was running out of coal and encountered a gale on July 30. A London *Times* reporter on board wrote: "If the wind lasted, we should be reduced to burning the masts, spars, and even the decks.... It seemed to be our particular ill-fortune to meet with head-on winds whichever way the ship's head was turned. On our journey out we had been delayed and obliged to consume an undue proportion of coal for want of an easterly wind, and now all our fuel was wanted *because* of one."[18]

On August 4, both ships after travelling half the distance across the Atlantic made landfall within hours of each other.

After contacting his wife, Field telegraphed the AP giving a brief description of the voyage and the White House alerting President James Buchanan that he would be receiving a message from Queen Victoria.

After the heads of state had their communication, the cable began to show its worth. On August 14, two overdue ships, the *Europa* and the *Arabia*, collided off Cape Race, Newfoundland. Ordinarily there would have been no news for friends and family, but with the cable, telegraph messages sent from Newfoundland alleviated their concerns with news that there was no loss of life. And the cable was now used for the transmission of foreign news. On August 27, the New York newspapers printed the following verbatim as it came over the wire: "News for America by Atlantic Cable:—Emperor of France returned to Paris, Saturday. King of Prussia too ill to visit Queen Victoria. Her Majesty returns to England, August 30. St. Petersburg, August 21st—Settlement of Chinese Question: Chinese Empire opened to trade; Christian religion allowed; foreign diplomatic agents admitted; indemnity to England and France.[19]

This transmission was an example of the value of the telegraph. The "Chinese Question" had occurred two months earlier with the signing of the Treaty of Tientsin. The information about the treaty had just reached England by ship. The telegraph reached the United States on the same day, taking two minutes for the complete transmission.

The jubilation over the laying of the cable was, however, short lived. Problems began arising in the transatlantic transmissions. The signals began to weaken and in some cases stopped altogether. On September 6, 1858, George Saward, secretary of the Atlantic Telegraph Company, sent the following letter to the London *Times*, "Owing to some cause not at present ascertained, but believed to arise from a fault existing in the cable at a point hitherto undiscovered, there have been no intelligible signals from Newfoundland since one o'clock Friday, the 3rd inst." Electricians "are investigating the cause of the stoppage with a view to remedying the existing difficulty. Under these circumstances, no time can be named at present for opening the wire to the public."[20]

By the end of September, it became public knowledge that the cable had failed. As a consequence, Field went from a hero to a scapegoat, and the public turned against him. As an example, Charles O'Conner, a distinguished New York lawyer, proclaimed "that this apparent sleight-of-hand stunt, a deception from start to finish, was a classic demonstration of how easily the public could be fooled."[21] The organizers of the company, having had problems before with the enterprise, simply went back to work to solve the problem.

The breakdown of the cable gave Samuel Morse the dismal satisfaction of feeling vindicated. He had prophesied that the cable could be laid and would work. "These points are successful," he said. What went wrong had nothing to do with his invention or his thinking, but with

5. The Atlantic Cable

the shabby motives and faithlessness of others. While abroad he learned that two Canadians had been named honorary directors of Atlantic Telegraph, neither of whom were stockholders—confirming his belief that he had been *"ejected"* from the expedition deliberately. His removal and the breakdown of the cable had the same meaning, and the same cause—the detestable mentality of trade that saw transatlantic telegraphy as a speculation, a *"money matter."* Mere money making "was the great and I might almost say exclusive motive of Mr. Field.... The hasty and unfair means he used to grasp too much, have resulted in utter failure."[22]

The investigation into the causes of the breakdown showed that there were problems in two areas: first, the company's chief engineer, Edward Whitehouse, had unknowingly destroyed the wire inside the cable by increasing the voltage in an effort to boost the transmission speed; and secondly, the gutta percha rubber cover on the cable had been damaged by being coiled and uncoiled when being moved in and out of the ships that were laying the cable. Also, the cable had been stored outdoors for a year before being used, causing the insulation to decay. "Take all these things together," Henry Field[23] wrote, "and the wonder is not that the cable failed after a month, but that it ever worked at all."[24]

These blunders cost the company $1,895,000 in addition to the loss of $140,000 in guaranteed annual income from the British and America governments because the cable failed. In May of 1859, Field again journeyed to England to try again—all he needed was more time and more money.

With the Atlantic Telegraph Company having its problems, another company, the North Atlantic Telegraph Company was founded by an American electrical engineer, Tal Shaffner, who believed that he could accomplish what Field could not. Shaffner believed that the 1858 Atlantic cable failed because it was in the wrong place. His cable would run from northern Scotland through the Faroe Islands to Iceland, from Iceland to Greenland and from Greenland across Davis Strait to Byron's Bay, Labrador. Shaffner put together an expedition to survey the route of his cable. The total run of cable would be 1,755 miles broken down into four parts, the longest being the distance of 700 miles between Iceland and Greenland. The electricians in the group liked the idea of the four shorter cables rather than one long one; the engineers, however, had a problem with the route because of the icebergs that formed along the coast of Greenland which they believed would crush the cable.

Though Shaffner's scheme seem to be an attractive alternative to Field's, especially with his "stepping-stone" idea, he could not raise the 500,000 pounds needed to begin working. Without the money, Shaffner had to give up, and he returned to the United States. There were also two other cable ideas at this time—one from southern Spain to Brazil and another from Portugal to South Carolina. Neither could gain any support at the time, although in the future both would be built. With the outbreak of the Civil War, a renewal of Field's cable attempts also came to a standstill.

Field did not give up. Realizing that his small group of investors could not fund another try, which would cost 600,000 pounds, he approached William Ewart Gladstone, the chancellor of the exchequer. Field asked for a guarantee of 8 percent interest for twenty-five years on new capital, regardless of success or failure. Fortunately for Field, England was still interested in the cable because they wished to be connected to Canada and because the United States had become an important maritime rival.

Because a cable across the Red Sea sponsored by the British failed, they were interested in doing business with Field. They raised the yearly subsidy to 20,000 pounds but would not give him the guarantee, and additionally, a committee from the British Board of Trade would have look at and approve any cable companies' technology. A new cable connecting Europe and India through the Persian Gulf was laid with many improvements over the Atlantic cable. This cable, which was the same length as the Atlantic cable, used half as much copper and twice as much iron for armor. It was a success.

During the Civil War Field traveled back and forth between New York and London looking for new sources of funding for another try at laying his cable. With the war in America, there was no interest or money for a cable investment until the Trent Affair.

On November 7, 1861, the British mail steamer *Trent* sailed from Havana with two Confederate commissioners on board. The ship was intercepted by a U.S. naval vessel and the two commissioners were taken off. This seizure was illegal since the men had been aboard a British ship. In case of hostilities, the British loaded 8,000 soldiers on the *Great*

5. The Atlantic Cable

Eastern and shipped them to Canada to guard the border. Secretary of State Seward defused the situation by stating that the American ship had acted without orders. The importance of the incident to Field was states in the words of the London *Times,* "We nearly went to war with America because we had not a telegraph across the Atlantic."[25] Diplomatic exchanges at this time took four to six weeks to cross the Atlantic.

Field now approached Major General George McClellan urging the general to connect all the forts on the Atlantic Coast between Washington and Florida by underwater telegraph. The general wrote back endorsing the idea and suggesting that Field be the one to implement it.

Armed with McClellan's letter and the London *Times* article, Field wrote to Seward about the cable and Seward agreed to see him. Field also asked for and was granted an audience with Lincoln. Although in a message to Congress in December of 1863, Lincoln recommended an Atlantic telegraph, he was more interested in a West Coast telegraph through Russian Alaska and then on to St. Petersburg. Lincoln distrusted the British, who were aiding the Confederacy. In addition to talking to Seward, Field also crossed the Atlantic to meet with Lord John Russell, the British foreign minister.

Once again, the British turned Field down, but he gained a new customer. Reuter's News Agency promised "no less than 5,000 pounds a year for the use of the line."[26]

Back in the United States, Field attended to the American part of his cable business, eventually purchasing all their important competitors on the Eastern Seaboard. By October 12, 1859, Field had reorganized the company to the extent that the only two large cable companies in the nation were the American Telegraph Company and the Western Union Telegraph Company.

Eight days later Field again returned to England, going directly to the offices of Glass, Elliot and Company, his cable manufacturer. Before he left England, Field had urged the company to undertake the cable project at their own expense by providing the wire, the men, and the ships, a proposal the company promised to consider. Glass, Elliot's answer was that for an estimated cost of 675,000 pounds, they would take full responsibility for the cable and manufacture it to the standards set forth by the Board of Trade. They would also undertake to connect all the Union forts along the Eastern Seaboard by undersea cable, if the United States wished to have it done.

On January 1, 1863, Field again returned to the United States to raise money, first from the American government and then from potential investors in New York City. The problem here was that the New York Chamber of Commerce had just passed a resolution condemning Great Britain for siding with the South in the war, and here was Field looking for money for a project that would benefit Great Britain. Gaining little support in New York, Field logged 1,500 miles, visiting cities on the East Coast looking for backers. By March he had commitments for 66,000 pounds, one-tenth of the capital needed. In June, Field once again returned to England to find that the money raising efforts had gone just as badly as in America. In August, Field returned home, having raised an additional 220,000 pounds; the total amount now raised was less than half of the needed funds. At this point, the people with large fortunes such as the Astors were avoiding him, leaving him with the reality that the only way that he could raise the remaining capital was by selling his American company, which he did in October of 1863.

Finally, in February of 1864 upon returning to England, Field's luck changed. He met with Thomas Brassey, the world's largest railroad builder, whose company employed 80,000 men on four continents. Field's recollection of the meeting: "In attempting to enlist him in our cause he put me through such a cross-examination as I had never before experienced. I thought I was in the witness box. He inquired of me the practicality of the scheme—what it would pay, and everything else connected with it; but before I left him, I had the pleasure of hearing him say that it was a great national enterprise that ought to be carried out, and, he added, I will be one of ten to find the money required for it."[27]

With Brassey's support, Field met with John Pender—the owner of Brett's Magnetic Telegraph Company, the company that would, in the future, supply all of the world's submarine telegraph cable—who agreed to match Brassey's contribution. The final investor was Daniel Gooch, chairman of Britain's Great Western Railway and majority owner of the huge *Great Eastern*, a vessel that Gooch could never find a use for that would justify its cost. These gentlemen subscribed all of the remaining funds needed for the venture. The last problem was the cable, which

5. The Atlantic Cable

was entirely redesigned. The new cable had seven small stranded copper conductor wires which formed the core. They were covered by four layers of gutta-percha rubber insulation, which was wrapped in tarred hemp, and protected by ten stranded wires wound in impregnated hemp. Field ordered 2,300 nautical miles of cable even though the surface distance from Ireland to Newfoundland was only 1,640 miles. There could be no instance where there was not enough cable.

The next requirement turned out to easy, which was transporting the wire across the Atlantic. Field knew that with the nation at war, the United States would not give him cable laying ships and in order to carry all this cable; he would need four ships which he doubted the British Admiralty would provide. Gooch provided the solution—the *Great Eastern*. With its gigantic size, it would carry all the cable and for free. If the expedition proved successful, however, the owners wanted 50,000 shares of Atlantic Telegraph stock. With the deal done, the cable was moved from Glass, Elliot, the manufacturer, to the *Great Eastern*. The amount of wire was so great that it took from February 1865 to June of that year to load the cable at a rate of twenty miles a day.

On July 15, the *Great Eastern* finally set sail carrying 7,000 tons of cable, 2,000 tons of water filled tanks, 8,500 tons of coal and 500 sailors, technicians and mechanics. Its displacement was five times that of the *Niagara*. The ship then left England sailing to Foilhummerum Bay on the west coast of Ireland, where the cable was attached to the companies' new telegraph station. On July 23, the *Great Eastern* again set sail, this time across the Atlantic. The next day after eighty-four miles of cable had been paid out, the communication between Ireland and the ship stopped. The electricians' instruments indicated that the problem was in the last ten miles of cable that was laid. All of that cable was brought back into the ship until the problem was found, a piece of wire had cut the cable. After splicing the cable, the *Great Eastern* resumed course only to have another problem almost immediately. After every expert on board could find no problem with the line, the decision was made to take up the cable again until the problem was found. At that moment, communication was resumed, although no one knew why. Again the *Great Eastern* continued its journey to Canada.

On July 29 the expedition was 634 miles from Ireland when again communication stopped. As before, the decision was made to bring in

How the Telegraph Changed the World

This page and opposite: A composite of scenes and people related to the laying of the Atlantic Telegraph cable, 1865 (Library of Congress).

5. The Atlantic Cable

and examine the cable that had just been released. Finally at 10 p.m. the break was found, the cable spliced and the voyage again underway.

The next morning the electricians examined the piece of cable that had been cut out and found that as previously happened what looked like a piece of wire had penetrated the cable exposing it to seawater, which cause communications to cease. The unanimous conclusion was sabotage, but who was the culprit? A cable had been sabotaged before on the underwater line connecting England to Holland by a rival company. In this case the Atlantic Telegraph Company only had one rival, Western Union, but all of the workers on board were trusted employees.

To avoid a repeat problem, volunteers were assigned to watch over the cable as it was uncoiled into the sea. On August 1, the *Great Eastern* passed the halfway mark without further incident and with the communication between the ship and Ireland working perfectly. But on August 2, the same problem reoccurred, a piece of wire was found to be sticking out of the cable. This time the wire did not penetrate the cable to its core and therefore communication did not cease and the electricians came to the conclusion that the problem was not sabotage, but rather a manufacturing defect by Glass, Elliot. Because of the change in wind direction coupled with the crew's efforts in bring up the cable, either the bow of the ship crossed over and cut the cable or the workers allowed too much strain on the line, and the cable snapped.

This time, instead of giving up and returning to England, the decision was made to attempt to fish up the cable from the seabed two miles deep. This had been done before in the Mediterranean at a depth of 700 fathoms, but never at this depth at mid ocean. The grapple (imagine a five prong fishhook) was played out and the quest for the broken cable began. Because of the weather, the ship had great difficultly locating the exact position where the break occurred. When it was finally found, a buoy was sent overboard with an anchor to the seabed, then fog rolled in making the buoy invisible. Finally, on August 7, the buoy was located and the fishing began. At 6 p.m. the cable was located and hooked, but the weight was too much for the machinery, and the cable broke again and fell back to the ocean bottom. A gale prevented a new search until August 10, when a new attempt was made and the cable hooked again, but the result was the same, the grappling hook could not hold the cable, which was now lost.

5. The Atlantic Cable

On August 17, the *Great Eastern* returned to England. John Russell, the writer for the London *Times*, filed a story of the attempt to lay the cable, which praised those who made the attempt, including Field, who was considered to be a hero by the British public. As usual, Field was undaunted by this latest failure, but amazingly, when he proposed to lay yet another cable, the Atlantic Telegraph Company was able to sell 120,000 shares of stock and immediately ordered new cable, making it clear to Glass, Elliot that their manufacturing defect caused the problem, without which the cable would be working. With the cable order placed, Field returned to the United States for his daughter's wedding, believing that the cable would be ready during the summer of 1866.

More problems arose, however; first, the Western Union Company had begun work on a landline through Alaska to connect with Russia, and second, the work on the new cable was progressing slowly, as a letter from Captain Anderson of the *Great Eastern* confirmed: "I am sorry you are not here. Somehow no one seems to push when you are absent."[28] Field immediately left for England to find that work on the cable had stopped and the shareholders' money returned to them because the issuance of the shares required the approval of Parliament, and Parliament would not meet until February.

Upon hearing this, Field went directly to his main financial backers—Gooch, Glass, and Brassey. Once again Brassey saved the day by committing to 10 percent of the cost of the next attempt even though his railroad construction company was in trouble. With Brassey as the mainstay, Field formed another cable company, the Anglo-American Telegraph Company, his fourth, and quickly gained the necessary financial subscribers. This time only 1,660 miles of cable would be required because there was still unused cable in the *Great Eastern*. The total left 113 miles of cable as a reserve.

By 1866, Field's one competitor, Western Union, had been making great strides in the West building their overland telegraph connection to Europe. Three hundred miles of wire had been strung across British Columbia to Alaska, another 350 in Siberia, and the Russians were pushing 7,000 miles of cable eastward from St. Petersburg. Laying the Atlantic cable now became an emergency.

In England, changes had been made to the cable which now had stronger and lighter galvanized wires, and the paying-out machine had been completely modified and strengthened. On July 13, 1866, the expedition once again left Ireland to lay the Atlantic cable. On this trip the only problem was that twice the cable kinked and the ship had to stop so that the cable could be uncoiled before it reached the paying-out machine. Finally on July 27, the *Great Eastern* reached Trinity Bay, Labrador, but as usual there was a problem. In this case, the Gulf of St. Lawrence cable failed connecting Labrador with New York. Field immediately undertook to repair the line, fearing that any delay would allow Western Union to complete the trans–Siberian line ahead of him. The repairs took two days, and on July 29, 1866, the AP received his telegraph message which read, "All well. Thank God, the cable is laid, and is in perfect working order."[29]

On the day the connection between Europe and America was completed, it carried messages at the speed of 7.36 words per minute. The world read closing quotations on Wall Street, prices on the Brussels grain market, and that Congress had finally readmitted President Andrew Johnson's state of Tennessee to the Union. When there was no traffic on the cable, each station called the other every fifteen minutes, using the code "NN," which meant "Nothing for you at present."[30]

Even with the success of the laying of the cable, Field was not satisfied, he wanted more lines across the Atlantic and the fastest way was by raising the cable that had broken off and was lying on the seabed. On September 2, after repeated attempts, the broken cable was recovered, spliced to the cable on the *Great Eastern,* and the ship then returned to Trinity Bay. On September 8, the second trans–Atlantic telegraph line became operational.

In May 1867, Captain Hamilton, skipper of the New Bedford whaling bark *Seabreeze,* pulled into a bay between the Amur River and the Bearing Strait in Siberia. Upon hearing of the arrival of a ship, George Kennan, quartermaster of Western Union's telegraph line across Russia, rowed out to the *Seabreeze.* Captain Hamilton asked what he and his men were doing and Kennan replied that they were building a telegraph line.

"A telegraph line," Captain Hamilton snorted. "Well if that ain't the craziest thing I ever heard of! Who's going to telegraph from here?" Kennan explained the project and then asked about the Atlantic cable.

"Oh, yes," the captain replied, "the cable is laid all right."

5. The Atlantic Cable

With a lump in his throat, Kennan asked, "Does it work?"

"Works like a snatch tackle," Hamilton replied. "The Frisco papers are publishing every morning the London news of the day before. I've got a lot of them on board I'll give you."

On both sides of the Pacific, work ceased. Russia eventually completed the project connecting Moscow with eastern Siberia, but as a result of the success of the Atlantic cable, Western Union lost $3 million.[31]

Cyrus Field, now forty-seven years of age, sold his interests in the four cable companies that he had stock in; also, for every share he owned in the American Telegraph Company, he now owned three in Western Union. In Great Britain, four of the men involved in the construction of the cable were knighted. The queen regretted that she could not knight Field since he was an American. After the celebrations ended Field discovered that he had debts going back to the old Field and Company. Settling these debts cost him $200,000, adding to the fact that laying the cable cost him his personal fortune. The cable itself turned out not to be a financial bonanza because it was too slow. It could only transmit fifteen words per minute as against four hundred words per minute on a landline.

Finally in 1873, Field's finances improved when he convinced the Anglo-American Telegraph Company to buy his 30,000 shares of the New York, Newfoundland, and London Telegraph Company at ninety dollars a share. Field never got out of the telegraph business and never stopped traveling between Europe and the United States.

At the time of Field's death in 1892, ten cables spanned the Atlantic, carrying an ever increasing load of messages to points around the world. "Columbus found one world and left it two. Cyrus W. Field found two continents and left them one."[32]

6

The Telegraph and the Railroads

When telegraph lines were first erected during the early 1840s, many were situated on poles alongside railroad lines, although the telegraph had no connection to the railways at that time. One of the practical problems that this situation caused was accidents. Generally, the mishaps came about because the poles leaned in towards the train and could strike the brakeman sitting atop the train, also because the telegraph wires could sag and catch on to a moving train. This resulted in the death of a passenger on the New Haven and Springfield Railroad in Connecticut.

When faced with these problems, the telegraph companies had a choice—they could establish a partnership with the railways as had been done in England and continue to have their lines on the railroad right-of-way, or move the lines to public highways. Without a partnership, the railroads began to look upon the telegraph as an instrument that interfered with their operations. At the same time, the telegraph companies were also looking at the advantage of leasing railroad routes for the telegraph. The following is a report from Francis O.J. Smith, a former congressman and president of the New York and Boston Magnetic Telegraph Association, to his board of directors:

> This seemingly endless catalogue of accidents and hindrances incident to the conjunction of telegraph lines with a railroad, and especially in this climate, with each under a distinctive administration, and of which an impatient public takes but little count in their strictures on telegraph lines, and the great losses consequent therefrom, illustrated to the directors of the line *the great error* involved in the preference hitherto given to railroads over public or country roads for the site of telegraphs, and induced their advice and resolution to authorize and hasten as much as possible, *the absolute removal* of at least one wire of their line.[1]

6. The Telegraph and the Railroads

This attitude persisted when Jeptha H. Wade, one of the founders of Western Union, attempted to build a telegraph line along the right-of-way of the Michigan Central Railroad. He said of the situation, "Railroad officials claimed they not only did not want it [the telegraph], but could not afford to have it along their road as it would endanger both trains and passengers, and as to it being of any benefit to them, that was all nonsense."

Wade said that J.W. Brooks, president of the Michigan Central Railroad, "met with ridicule my arguments to convince him that [the telegraph] could be made useful in railroad business, and said he was surprised that a sensible man like me should make such a claim to one of his experience." Brooks said to him, "Why, I had rather have one hand car for keeping my road in repair and handling my trains than all the telegraph lines you can build."[2]

During the 1830s and 1840s in the United States, there was little reason to believe that the telegraph could be of any benefit to the nation's railroads, even given the experience of the British. In the winter of 1849–50, a conductor on the Michigan Southern Railroad telegraphed ahead to hold a boat for the passengers on his train. This is probably the first instance of a telegraphed train order in America. There is no doubt that the English were the first to adopt telegraphic signaling on railways. A pamphlet entitled "Telegraphic Railways; or, the Single Way Recommended by Safety, Economy and Efficiency, under the Safeguard and Control of the Electric Telegraph, &c. By William Fothergill Cooke, Esq.," published in London in 1842, has a large chart illustrating fully the manner in which trains were to be moved on a single track by means of telegraphic signals or orders, given by the station masters from station to station. As an example the chart also depicts in diagrams the telegraph system used on the Blackwall Railway and gives explanations for the use of each piece of apparatus.[3]

By the late 1840s, some long distance rail lines in the United States began using the telegraph as a means of dispatching its trains, but not the shorter distance railways, whose management still believed that the telegraph apparatus was dangerous because they did not control it. Added to that, the smaller lines did not have the financial resources to build telegraph systems.

Beginning in the 1850s more and more railroad lines began using the telegraph as an adjunct to their usual means of operation—the use of timetables and strict operating rules for guiding train movements. The Pennsylvania Railroad, in its annual report dated January 1852, indicated that the railroad managers "found telegraphic communications to be a helpful, but not an essential, aspect of operational management."[4] The usual method of operation was that after station agents closed their telegraph offices for the night, the managers expected the train crews to depend on the old and more fundamental tools of railroad management such as schedules that had been developed in the previous decade. The Pennsylvania line had, however, expanded its use of the telegraph in 1849, when the Atlantic and Ohio Telegraph Company built a line alongside the railroad's entire length in exchange for doing all of the railroad's telegraphy for free.

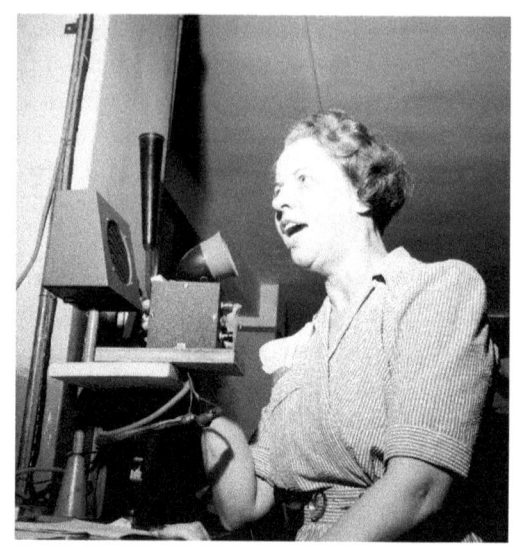

An operator talks into a microphone of the office intercommunication system at Western Union telegraph office, 1943 (Library of Congress).

In 1851, the New York and Erie Railroad Company, a competitor of the Pennsylvania Railroad, became the first railroad in the nation to dispatch trains telegraphically. However, the telegraph was only to be used when excess seasonal traffic caused delays and backups on the rail line. "The telegraph is only to permitted to be used for [train dispatching] when the trains have become deranged, and then only by one person, on each division, specially authorized to perform this duty."[5]

The difference between the Erie and the Pennsylvania railroads as far as the telegraph is concerned was one of control. The Erie owned its telegraph line, the Pennsylvania did not and furthermore could not operate it when it was being used by the Atlantic and Ohio Company.

6. The Telegraph and the Railroads

While the use of the telegraph during the Civil War is covered in a separate chapter, it is worth noting that by the end of the war, the United States Military Railroad combined with the Military Telegraph Corps had provided valuable communications and logistical support for the Federal forces. It was not the same at the beginning of the conflict, however, when few military officers had any useful knowledge of either railroads or the telegraph. With the end of the war, even with its proven successes, the combination of the two did not engender much interest as a civilian application. Perhaps the reason was the opinion of General Herman Haupt, Lincoln's railroad chief who vocally derided the telegraph as unreliable and ill-suited for railroad work.

With peacetime railroad traffic began to increase to the point where many railroad managers throughout the East and Midwest evaluated the economic and organizational costs and benefits of telegraphic dispatching and concluded that it was necessary for coping with these new operating demands. Few rail lines, however, had the expertise or necessary financial resources to install a telegraph system. Consequently many railroads continued to run trains under poorly developed wartime operating practices instead of taking advantage of a better and safer system.[6]

After the war, the railroads in the Northeast were almost unique in their opposition to using the telegraph to dispatch their trains on a constant basis. The train lines in this section of the country were basically unaffected by the war and therefore were not forced to make any changes in their operations. They believed that the systems they had in place were superior to any involving the telegraph, and, they had already faced heavy volume situations before the war and therefore believed that the systems in place could manage peacetime traffic increases. Finally, these lines had had no problems operating on a fixed schedule.

One company, the Stonington Line of Connecticut, even used avoidance of the telegraph as a benefit in their advertising. "No Trains Run by Telegraph," their ads ran on a mid–1880s placard, along with ad copy reading, "Reclining Chair Cars on all Express Trains." Obliviously, the line believed that its customers mistrusted the telegraph or was trying to turn a negative (not having the telegraph) into a positive.

The first railroads in the United States were constructed on a single line of track. Head on collisions between locomotives were a major hazard. This meant that the movement of trains had to be carefully controlled if goods and passengers were to be moved with safety and a degree of efficiency. The early optical telegraph (semaphore) was able to provide a degree of control, but was limited by the weather and darkness. The railroad building boom that began in the 1840s only intensified the need for more efficient communication in the railroad industry. The invention of the telegraph in 1838 and its subsequent adoption for commercial use in 1847 provided an answer to the control needs of the railroads.[7]

On August 26, 1871, a horrific accident occurred on the Eastern Railroad of Massachusetts that turned public opinion toward a better method of train scheduling. This particular accident could have been avoided had the line used the telegraph which was available, but seldom used. Charles Francis Adams, the historian, politician, diplomat, writer, and president of the Union Pacific Railroad said of the accident, "It may with perfect truth be said that the disaster at Revere marked an epoch in the history of railroad development in New England. At the moment, it called forth the deepest expression of horror and indignation, which, as usual in such cases, was more noticeable for its force than for its wisdom."[8]

Adams continues, regarding the telegraph, "There is nothing new or experimental about it. It is a system which has been forced on the more crowded lines of the world as an alternative to perennial killings.... This opinion was expressed also after the Revere disaster of 1871. It should have been branded into the record of the state the impossibility of safely running any crowded railroad in a reliance upon a schedule [alone]. Such men as this however are not accessible to argument or the teachings of experience, and the gentle stimulant of a criminal prosecution seems to be the only thing left."[9]

As the expansion of the telegraph was being considered by the railroads, a new idea came to the fore to help avoid accidents. This was the block system which would be used in conjunction with the telegraph. The system operated as follows: a train would enter a stretch of track (a block) with a signal tower at its entrance. A signalman would telegraph ahead that a train had entered the block but would not be allowed to proceed until an operator at the other end of the block signaled back that the block was empty. This new system ensured that delays on busy lines would not lead to rear-end collisions.[10]

6. The Telegraph and the Railroads

Adams championed the new system because even though he supported the telegraph, telegraphic dispatching alone only provided the illusion of safety, since collisions could still occur on telegraphically dispatched rail lines due to employee error or unexpected train delays. The problem for the railroads in choosing this system over depending on scheduling alone was cost. As an example, the capital outlay for the establishment of the block system on the eighty-nine miles of track between Jersey City and Philadelphia would have been over $20,000 for the telegraphic equipment, plus the cost of the block towers. Additionally, wages would have amounted to nearly $2,000 per month.[11] Few companies in the early 1870s, with the exception of the Pennsylvania Railroad, could have afforded the block system.

This lack of finances was the basic problem for American railroads during the post war era. It was especially true concerning spending money for accident prevention. With the block system financially unavailable, the best that the rail lines could do was to continue using the telegraph and the other antiquated systems such as relying on scheduling. Moreover, many lines could not even afford installing a telegraphy system. The beneficiary of this problem was Western Union. Instead of buying the telegraph apparatus, these lines leased telegraphic service from the telegraph giant. While this service was inexpensive, it was also given on a shared basis, which prevented managers from issuing orders in a timely and efficient manner. The result was that concerns about operating costs won out over concerns about safety, but more than that it allowed Western Union to exercise a great deal of control over American railroads and to reinforce the telegraph companies' inertia concerning technological advancement.

Samuel Morse became involved in looking at railroad safety when a director of the St. Germain railroad, a twelve mile line in the western part of Paris, asked him whether a telegraph could be contrived to report the location of the company's trains along the road. Morse worked out a detailed plan featuring small electromagnetically operated bell towers that would register the arrival of the cars at a given point and notify every station over the whole route. He took a brevet[12] for his new invention and made a working model.

The directors of the railroad, however, failed several times to show up for the demonstrations Morse arranged at his lodgings. "They are

How the Telegraph Changed the World

famous here for not keeping appointments," Morse said at the time, "I have only to exercise patience and wait." He needed more than patience. When the directors finally came they approved his plan—all but one of them, who objected to the enormous cost of building a trial line, an estimated 60,000 francs. Morse proposed a less expensive set-up: instead of stretching the wire circuit underground, in protective tubes, it would be laid above ground in grooved bricks. He waited for a reply, but heard nothing. As far as Morse was concerned, like the rest of France, the railroad officials proved to be "as dilatory as the Government." In the end the company simply dropped the project.[13]

In 1872, "Hindoo," a British colonial official visiting the United States, wrote a series of articles for the *Railway Gazette*, one of the key trade journals for the American railroad industry. In an article on February 3, "Hindoo" looked at the "American System" of train dispatching in which a single dispatcher oversaw all the freight and passenger movements within his division and compared it to the train movement situation in India. Under the Indian system (borrowed from the British), before a train could leave a station, the telegraph operator at that station had to receive word from the operator in the next station that the line in between was clear. The most important feature of this arrangement, "Hindoo" noted, was that it eliminated the possibility that the dispatcher or other employees could make errors and cause collisions. While many railroad officials in the United States agreed with "Hindoo," the problem—as with any arrangement in which a telegraph system was purchased—was the cost.[14]

One of the railroad men responsible for using the telegraph in a cost effective and safety producing manner was Daniel C. McCallum, a general superintendent of the Erie Railroad. He stressed that channels of authority and responsibility were channels of communication. He paid close attention to improving the accuracy of information and the regularity and speed with which it flowed through these channels. Hourly, daily, and monthly reports were extremely detailed. The hourly reports, primarily operational and sent by telegraph, gave the location of trains and reasons for any delays or mishaps. "The information being edited as fast as received, on a convenient tabular forms, shows at a glance, the position and progress of trains, in both directions on every Division of the Road." Just as important, the information generated on

6. The Telegraph and the Railroads

these tabular forms was filed away to provide an excellent source of operational information which, among other things, was useful in determining and eliminating "causes of delay." McCallum's use of the telegraph brought universal praise from the railroad world both in this country and abroad. What impressed other railroad managers was that McCallum saw at once that the telegraph was more than merely a means to make train movements safe. It was a device to assure more effective coordination and evaluation of the operating units under his command.[15]

In spite of the problems associated with safety, the growth of the railroad was intimately tied to the growth of telegraphy. They were complementary technologies, each meeting the particular needs of the other, and in the case of the telegraph, enabling it to become a monopoly. Among the most important reason for this growth on the part of both were the vast area which comprised the United States, the needs of commerce, and the needs of politics. All of these encouraged the invention of a set of machines designed, in one way or another, to facilitate communication.[16]

"The electric telegraph ... was the first large-scale and commercially important use of electricity."[17] The railway companies were the prime users of telegraphy. These two industries developed hand and hand, spreading a vast transportation and communications network over the entire country.

> The railroad permitted a rapid increase in the speed and decrease in the cost of long-distance, written communication; while the invention of the telegraph created an even greater transformation by making possible almost instantaneous communication at great distances. The railroad and the telegraph marched across the country in unison. As has been pointed out, the telegraph companies used the railroad for their rights-of-way, and the railroad used the services of the telegraph to coordinate the flow of trains and traffic ... many of the first telegraph companies were subsidiaries of railroads, formed to carry out this essential operating service.[18]

The marriage of these two industries "went on until the whole railroad system of the continent has become more or less identified with telegraph interest, and with its whole transportation service regulated by its control."[19] Looking at the early contracts between the railroad and the telegraph companies, it becomes easy to see the mutual benefits each received from the other. Going further, it is difficult to ascertain which industry benefited the most: "To the railroad companies it is of inestimable value. Not only does it give them an economical means of control, and a share of telegraphic revenues, but the executive officers are usually provided with such wide facilities for communication with other companies and with their own officers at distant places, that the whole machinery of management has become sublimely prompt, consolidated and simple. To the telegraph company it is protection, economy, permanence, and strength."[20]

The following is a sample contract between a telegraph company and a railroad company:

> The telegraph company to furnish the railroad company with a single wire of proper size and quality, and provide Morse instruments at certain specified stations on the line of their road.
>
> The telegraph company to maintain main battery for working said wire day and night.
>
> The telegraph company to keep the wire erected for the railway company in order, except as otherwise provided.
>
> All receipts for messages at offices opened on the line of the railroad, by either party, to belong to the telegraph company.
>
> The railroad company not to send any message free except for its own agents on its own business.
>
> At all stations in addition to those named, the railroad company to supply all machinery and local battery.
>
> The railroad company to instruct its men to watch the line, straighten poles, reset the same when down, mend wires, and report to the telegraph company.
>
> The railroad company to convey and distribute wire and insulators and all other material free, and also furnish a hand car for stringing wire.
>
> The two companies to reciprocate the use of wires when those of either are out of order, but railroad wire never to be interrupted when sending railroad business.
>
> The railroad company to transport all instruments, material for repairs, all operators, officers and agents of the telegraph company free of charge when on business of the company, and to furnish and distribute poles when line has to be renewed, the telegraph company setting and insulating the same.
>
> The railroad company to pay for stringing the railroad wire and insulating the same, and for instruments, etc., thirty dollars per mile.
>
> The railroad company not to allow any other telegraph company to build a telegraph line upon its property.
>
> Railroad telegraph operators may accept public business at the ordinary tariffs, and shall account for the same to the telegraph company, but no messages will be accepted or sent to interfere with railroad business.

6. The Telegraph and the Railroads

Of course the terms of these contracts varied with circumstances, but they were all liberal. They bear the evidence of two parties whose interests were mutual, cordially and liberally providing for each other's necessities. Thus the rail and the wire became indissoluble, and have carried civilization and civilizing influences wherever they have gone.[21]

Western Union took full advantage of the section of the contract that stated, "The railroad company not to allow any other telegraph company to build a telegraph line upon its property." Once its lines along the right-of-way were in place, Western Union controlled all of the railroad telegraph business, and since Western Union was the largest telegraph company, this clause effectively eliminated any competition. Consequently they were able to charge cheaper rates than their competitors which in turn added to their business and ultimately led to their position as the first modern business monopoly.

The fusion of railroad and telegraph technology allowed for the emergence of "strong and aggressive communication and transportation systems which transformed the life of the nation."[22] Alfred Candler went further, stating that the real system building, that is creating a structure which assured a "continuing flow of freight and passengers across the roads' facilities by fully controlling connections with major sources of traffic," did not occur in the railroad industry until the railroads had been integrated into a single national network. Consequently, the control made possible by the rapid communication that the telegraph provided did not achieve its fullest effect until the late 1880s when the other economic vagaries were dealt with (for example, effects of competing companies, managerial issues and the impact of small entrepreneurial enterprises—which effectively disappeared by 1900).[23]

7

The Telegraph and Business

In the year 1847, when the commercial telegraph was almost a year old, it was seen as "facilitating Human Intercourse and producing Harmony among Men and Nations.... [I]t may be regarded as an important element in Moral Progress."[1] "The telegraph system is invaluable," a business journalist declared twenty years later, "and when the missing links shall have been completed of the great chain that will bring all civilized nations into instantaneous communication with each other, it will also be found to be the most potent of all the means of civilization, and the most effective in breaking down the barriers of evil prejudice and custom that interfere with the universal exchange of commodities."[2] After seeing its uses in war, *The Commercial and Financial Chronicle* foresaw improvements "upon the political atmosphere of the nations thus brought into closer moral contact with each other," and a decline in "international hostility...[3] The hand of progress beckons unceasingly to freedom, and whenever science achieves a victory, a rivet is loosened from the chains of the oppressed."

After its invention and when its uses became apparent, in the world of business and finance, the telegraph was regarded as an agent for expanding competition. Market information including the prices of goods and quantities of product available would now be available to everyone on an equal basis; monopoly power would be weakened. "The telegraph being alike open to *all* puts the whole community upon a par, and will thus '*head off*' the most adroit speculators, because they will not have the power to *monopolize* intelligence," a Philadelphia business writer predicted in 1846.[4] Everything about the "lightning wires" appeared to favor small business enterprises.

Initially, this was the case, the telegraph did bring about reductions in regional price differences, as well as lowering the costs of gathering and distributing information. For businesses, it saved time, cut down

7. The Telegraph and Business

on the need for large inventories, decreased short-term financing requirements, and in some cases allowed for the elimination of middlemen and wholesalers. It assured equal access to commodity and financial markets and, coupled with the railroad, turned regional markets into national ones.

These national markets, however, were not necessarily competitive markets. In point of fact the telegraph industry became the first American monopoly in 1866 when Western Union bought out its last two rivals. More than that, by 1868 the United States and Canada were the only countries in which the telegraph was not under government control.

Before the telegraph, the expansion of business was curtailed by the cost of selling a product outside one's immediate market. With the telegraph and its long distance instantaneous communication, the cost of doing business plummeted. The average size of firms began to increase as business increased because of companies' expanded markets. As an example, the telegraph enabled businessmen to "speak to" a correspondent or banker in another city "and get a response, in the same hour."[5] In 1849, Philadelphia merchants were promised "the business of the far west, and a mass of business which New Orleans must pour in [once] the union of the lines from Philadelphia to Louisville should be consummated."[6] By the early 1860s, "The volume of sales which a single firm handled jumped from an annual volume of tens and hundreds of thousands of dollars to tens of millions of dollars. Data on wholesalers in cities other than New York suggest that as soon as they reached out for the markets of the hinterland, they became as large as any mercantile enterprises in history, with annual sales figures of $10 to $50 million and higher."[7]

No longer did storekeepers of the South and West have to make their semiannual treks to the eastern markets. Now he could telegraph his needs to a wholesaler, who also benefited from the telegraph because he did not have to carry the large inventories that he did in the past and because he could order directly from the manufacturer by wire and be fairly certain of delivery on a specified schedule.

The telegraph also allowed for a move of the wholesalers from the Eastern Seaboard west to open new markets, as an example—Cincinnati in 1859:

How the Telegraph Changed the World

> Within the last eight or ten years Cincinnati has been gaining a position as a great centre of supply by wholesale, to country merchants of Ohio, Indiana, Illinois and Kentucky, of their dry goods, groceries, hardware, boots and shoes, hats, drugs, and fancy goods. In those various lines of business it is becoming very apparent to purchasers that they can deal here to greater advantage than our eastern cities. The effect of this has been to enlarge our sales to country merchants. For example—dry goods, from $4,000,000 in 1840 to $10,000,000, in 1850, and to $15,000,000 at this time. There is a corresponding increase, also, in all other descriptions of business which go to make up general sales to country merchants.[8]

Paralleling this development of the wholesale business was that of the commodity exchanges, for the same reason—electronic communication. Between 1845 and 1854 exchanges for wheat, corn, oats, and cotton were established in Buffalo (1845), Chicago (1848), Toledo (1849), New York (1850), St. Louis (1854), Philadelphia (1854), and Milwaukee (1854).[9] Their dependence on the telegraph was unmistakable, and from the start they "were almost wholly devoted to the [national] grain trade rather than the general commercial interests of the cities in which they were located."[10]

The first of the mass retailers, the department stores, appeared in the 1860s and 1870s in cities where populations became large enough to support them. These stores—Macy's, Arnold Constable, Marshall Field, and Wanamaker's—depended not only on population, but also on a fast turnover of inventory for their consumer goods. The telegraph was the key to effective inventory control, which gave them a competitive edge. Meat packers also made use of the ability to communicate quickly. Their head offices were "in constant telegraphic communication with the branch houses and commission agents during the process of the sale of each carload of beef, obtaining information and giving advice." Remote control was imperative in a nationwide business based on bulk purchases of cattle, small margin between cost and selling price, and prompt sales at specific times to avoid spoilage.[11] In the cotton trade following the Civil War, the telegraph was employed by a new class of dealers to purchase cotton from planters, farmers, and storekeepers at the railheads and to sell directly to mills in New England and Europe. Although the telegraph in essence decentralized cotton marketing, moving it away from the few southern port cities where the sales were basically handled by middlemen—factors—and relocating it inland to a number of small

7. The Telegraph and Business

markets, the net effect was to concentrate the buying of cotton in the hands of strategically placed cotton dealers before the production and distribution process. Between 1870 and 1900, a small number of cotton merchant firms, American and European, came to dominate the business, transferring cotton by telegraphic orders throughout the world.[12]

After the Civil War, one of the ways that American corporations became larger was by merging. These firms remained profitable only if, after consolidating, they adopted a strategy of vertical integration. Two points need to be kept in mind. First, mergers on a national scale appeared only as the railroad and telegraphic network went into full operation in the 1870s and 1880s. By lowering transportation barriers, the railroad permitted many small enterprises to compete in the national market for the first time. And simultaneously, the telegraph helped to make possible centralized supervision of a number of geographically scattered operating units.[13]

For private enterprise, the attraction of the telegraph lay in its unmatched speed of communication, and speed was relatively more important for business than any off-setting cost of its telegraph service.[14] Communication technology in the nineteenth century promoted large-scale bureaucratic organization at least as much as it promoted better functioning of markets.

"It was the telegraph which first brought the speed of electronic communication within reach of the potential empire builder."[15] The "imperial" features of the telegraph in business operations were prominent from the outset. Not only could employees be controlled and geographically scattered units brought under centralized supervision, but immediate transmission of orders and requests was now assured. As with the military during the war, secrecy for the users was guaranteed through the use of elaborate codes, as well as by strict telegraph company rules, designed to ensure strict confidentially of the messages. Even as early as 1844 and 1845, codes were used to disguise the messages that were sent. For instance, "cotton is on the rise" and "stocks have fallen" were rendered by one company as "Zqoq&wfq&ots" and "foqzuftrgsxrvs&."[16] Private telegraphic codes were employed by several large firms, the Du Pont Powder Company had a code manual that grew to 600 pages by the turn of the century with phrases like "adermennig" ("carload not ready") and "habitadme" ("how many were killed?"). Its distribution was restricted to 305 employees in the period 1901–1905.[17] The telegraph also helped employers combat unions by allowing speedy circulation of blacklists of known labor organizers.[18]

The intelligence value of the telegraph was augmented by private telegraph lines. In 1849, R.M. Hoe, a director of the Magnetic Telegraph Company, had a line strung from his New York City telegraph office to his printing equipment factory in the same city. "Its value was at once evident. The master was felt to be among his men although two miles intervened."[19] This idea of an intercity telegraph was first tried in Philadelphia after the war. The Philadelphia Local Telegraph Company created what became known as the "Telegraph Exchange" which connected local businesses telegraphically through a central office, so that they could conduct business directly with each other or with the central station. A similar company was incorporated in New York in 1869 to establish telegraphic communication between all parts of Manhattan Island. This organization, the Metropolitan Telegraph Company of New York, charged a uniform price of 10 cents per message. In spite of this low price and ample advertising, the venture failed.

Another idea that did not fail began in 1863 when S.S. Laws, a presiding officer of the Gold Exchange in New York, developed an instrument that gave the price of gold when the market opened. Since at this time, the opening price of gold regulated the day's prices on a large variety of products, every morning great crowds gathered outside the exchange to catch the first figures from the gold indicator and then returned to their businesses to price their goods. Laws' idea was to relay this information electronically by means of the telegraph. So successful was this idea that Laws resigned from the Gold Exchange to create the Gold Reporting Telegraph, which gave the new price of gold with every fluctuation. Subscribers to his company were thus able to know to price of gold on a continuous basis.

With the addition of high speed printers between 1868 and 1880, private telegraph lines spread even more quickly. One of these, the Calahan printer, known more familiarly as the stock ticker because of the sound that it made, allowed the Gold and Stock Telegraph Company organized in 1867 to supply stock quotations for banks and brokerage houses in New York City. After this venture, the company began fur-

7. The Telegraph and Business

nishing complete telegraph lines, type-printing instruments, batteries, and trained operators "at a very moderate annual rental."[20]

By 1878 Gold and Stock, now in a market sharing "cooperative union" with the Western Union Company, controlled all the fast "stock printers" of proven commercial feasibility and was renting and servicing 300 private lines with 1,200 miles of wire in and around New York City with underwater cables connecting New Jersey and Long Island. Going farther, it expanded to include Cleveland, Baltimore, Louisville, Cincinnati, Buffalo, and Chicago.[21] In 1884, Western Union started leasing its own wires to brokers and large retail establishments, as well as to banks and press associations. During certain hours these arrangements allowed the lessees to employ their own operators and enjoy exclusive use of the wires. By the mid–1880s, leasing of wires became "so much more profitable than handling messages that the company had considered a suggestion that it cease to handle messages entirely and turn its entire attention to leased wire business."[22]

Even with all its benefits however, the telegraph could cause problems depending on the motives of its users. Newspaper dispatches of 1846–50, the opening years of commercial telegraphy, were laden with reports of the feeding of false price and quantity information by "speculators" over the wires. According to a New Orleans newspaper in 1847, Crescent City merchants had been so badly victimized by seemingly deliberate doctoring of telegraphed commercial and financial reports that they expressed as "their most fervent wish, that the telegraph may never approach us any nearer than it is at present."[23]

An example of the results of the spread of misinformation came in 1857. Obviously the first investor to acquire information, if accurate, gains a substantial advantage. "During the week of financial excitement, in October, the exaggerated reports of which were carried with the speed of lightening to every part of the land, this new medium of communication (the telegraph) filled our banks with imperative orders for the immediate return of their deposits, in specie," James Gibbons observed in 1859.[24]

American economist David Wells observed in 1889 that while the telegraph had brought "command ... of instantaneous information throughout the world of the conditions and prospects of all markets for all commodities" and had imparted a "steadiness to prices," it was also

How the Telegraph Changed the World

The grain exchange at Minneapolis, Minnesota, shows the telegraph office on the left, 1939 (Library of Congress).

"spreading in a brief time the same hopes and fears over the whole civilized world [and] made it impossible any longer to confine the speculative spirit to any one country."[25]

When the telegraph service began to connect cities, it started with the larger ones: New York, Philadelphia, Boston, Washington, New Orleans, and St. Louis. Telegraph users in smaller cities and towns often found that the lines were tied up by heavy volumes of messages traveling between and within major cities. The situation was aggravated by the priority status given to government, police, and some press dispatches. As an example, in 1848, a business reply sent in New York State from Troy to Rochester took more than twenty-four hours to be transmitted "solely because the [Troy] operator could not get his turn at the wires." A dozen attempts by the Rochester telegraph office to inquire about the status of the message met with frustration: "In the morning, two or three of the most important stations, such as New York, Albany and Buffalo, have exclusive possession of the wires till their business is completed [and this leads to] a struggle for the wires." In addition, there were peri-

7. The Telegraph and Business

odic breakdowns in service, a continuing "general scramble among the operators for the wires," and top priority "private messages for the President of the company." Before these were finished," the Rochester *Monroe Democrat* complained, "Troy has gone to bed."[26]

These problems could be fixed, however, when the benefits of doing so were made known to the company involved. In a letter to the president and directors of the Atlantic and Ohio Telegraph Company, James D. Reid, then a superintendent of the Magnetic Telegraph Company, stated, "I need not say that, thus limited, your line is utterly inadequate to public necessities. With proper facilities for communication, the business of your line must largely increase; but whether this were to follow or not, the public will demand of you the prompt transaction of their business, or give it to others, who may provide them with advantages denied by you." In his book *The Telegraph in America*, Reid wrote that in this instance, "I ought to add that my appeal was at once responded to, and the patentee reserved fund was ordered to be used for a new wire...."[27]

The basic problems were first, that commercial messages took up more time than private communications, and since they could contain monetary information, had to be relayed exactly; therefore, they required repetition and verification. Second, waiting time for any telegraph station depended on the number of messages being transmitted by "higher order" stations ahead of it, an advantage which busier stations passed on to their best customers. And finally, during the antebellum period, demand on the lines was unevenly distributed, with a small amount of urban stations generating 65 to 80 percent of total company revenues.[28]

8

The Telegraph and the Press

The expense of long distance news reporting, coupled with the public's desire for rapid and timely information in the antebellum United States, led newspaper editors to consider some sort of alliance among themselves. At this time the fastest means of disseminating information was the mail. The editors came to the conclusion that an association coupled with the telegraph would be the answer to the problem of gaining instantaneous news reporting.

By 1848, two such associations were formed in New York City: the Harbor News Association, dedicated to obtaining foreign news, and the New York Associated Press, an organization concerned with gathering domestic news. Subsequently, other associations were formed regionally: the New England Association, the Southern Association, the Western Association, and the New York State Association. Eventually all of these groups coalesced into what became known as the Associated Press.[1]

That telegraph companies entered into alliances with newspapers suggests that in addition to disseminating information, the companies involved would have a "monopoly at once powerful and profitable." This monopoly did not come at once because of the struggles for control of the information between the largest telegraph company, Western Union, and the largest press combine, the Associated Press (AP). By the 1860s, however, the two companies had a firm alliance through which they dominated domestic wire news services. As a result, this control over information allowed the AP to become an extremely powerful news agency. This alliance was alluded to in the 1870s by opponents of this type of control of an industry, leading to the establishment of the Postal Telegraph Company.[2]

The contract between Western Union and the AP was believed to be against public policy because it stipulated that the AP members would not "advocate the establishment of competing lines to the injury of the

8. The Telegraph and the Press

business of the Western Union Telegraph Company."[3] In 1866, Elihu Washburn, a congressman from Illinois, called for the government to construct telegraph lines in direct competition with Western Union, implying that under the present system, Western Union would establish favorable rates for the AP and less favorable rates for non-members. The way the Associated Press was said to control the establishment of new newspapers is through membership in its organization. In this way all of the fees associated with the transmission of data would be split among the members. An independent newspaper would have to bear the entire cost, where the members would only pay a fraction. The president of Western Union, William Orton, in his testimony, denied these accusations. He said that the telegraph lines were open to all papers, that there was no favoritism—simply an exercise of business. Western Union was interested in getting as much revenue as possible for its service and the newspapers were interested in paying as little as possible; therefore arrangements were made between the companies. After this congressional hearing, the topic of a monopoly between Western Union and the AP officially died. It continued, however, because of the collusion between the two companies. As an example, in 1894, the International Telegraphic Union complained that no new newspapers could be established without joining the association because of its ability to withhold news dispatches.

Along with the struggles between Western Union and the AP on the one hand and the federal government on the other, there was also a struggle among the various associations that made up AP, which culminated in the press association war of 1866–1867.

The press war situated around the New York Associated Press, the strongest component of the AP, and the other regional press associations. The New York Associated Press was the strongest part of the system because of its location and because of the telegraph. By 1846, telegraph lines spread east of the city to Boston, and north and west to Albany and Buffalo. The southern line to Philadelphia and Washington was in the process of extension southward along the seaboard, toward New Orleans. The telegraph also began to move West via Philadelphia and Pittsburgh connecting the Mississippi Valley with the Atlantic Coast centers. As a result, New York became the "one grand center" of this network.[4]

How the Telegraph Changed the World

Each of these routes was significant from the standpoint of American "news flow." War news (Mexican War) and congressional reports came from the South, steamer news from the East, and political reports from the state capital in the North. News items from every part of the country using telegraph lines could be transmitted directly to New York, and conversely news accumulated there could be "broadcast" throughout the country. The New York press for the first time was using a technological system (the telegraph), giving it both a national and international (the Atlantic Cable) field. The military historian Martin Van Creveld speaks of the combination of technology and information in what he calls "the age of systems":

> It was not so much that the newly invented tools and machines were more effective than their predecessors, though that was important, as the fact that they could only be used in integrated systems which did the trick. As more and more individual tools and machines surrendered to the drive for efficiency and were integrated into systems, success ... tended to depend on nothing as much as the ability to understand these systems and manage them.... The system approach [was] first made manifest in the telegraph and the railway.[5]

With the transformation of American telegraphic lines into a system in the summer of 1846, it became apparent that this type of organization was required to reap the full benefits of Morse's invention. Further, the combination of the telegraph with the projected reach of the New York newspaper allowed for the formation of the New York Associated Press.

The cost savings was another reason for the emergence of the association. While it was more expensive to use the telegraph to disseminate information than by using the mail, each New York newspaper would only have had to spend $50,000 to cover the cost of the entire American telegraph system in 1846. As a comparison, twenty years earlier, before the telegraph, the *Journal of Commerce* and the *Courier and Enquirer*, with circulations of about 5000 each, spent the same amount yearly on their express news arrangements alone.

The basic problem for the New York newspapers was therefore not cost, but rather competition. Before, the telegraph reporters from various newspapers vied with each other to provide their papers with scoops. Now, with the telegraph, editors complained that everyone would receive

8. The Telegraph and the Press

the information simultaneously. Samuel Morse's answer was the "first and essential principle of telegraphy"—that the first to come should be the first to be served. However, since there was only one telegraph line, Morse's answer would not do. The result was that reporters now lined up at the telegraph office waiting for news transmissions.

The seriousness of the situation of having only one telegraph line came to the fore in 1846 when President James K. Polk telegraphed information pertaining to the war in Mexico. The cities in the Mississippi Valley from Minnesota to Louisiana received the last words of the message forty-eight hours after it was sent. Part of the reason was weather related, but a major reason was the fact that the superintendent of the telegraph line, James D. Reid, held up the president's words so that other messages that had accumulated could be sent. He explained to the outraged press:

> A merchant in Cincinnati wanted to pay his note, due that day in New York. His dispatch must be delivered before 3 o'clock. At 1 [o'clock] it was on our files unsent. A mother was dying; she wished to see her boy—her eldest, her best beloved—before she yielded up the panting breath that struggled to be free. Her message was there too. A merchant, his goods seized, the hammer of the auctioneer ready for the sacrifice, appealed to his distant friend for aid. His message is there unsent. A hundred other of various concerns were thus grouped, the expressions of many hearts trembling, hoping, fearing the result of their distant mission. An hour or two, and these would have failed of their purpose ... and for what? For a message justly interesting, but involving to none living an individual joy, or advantage, or regret?[6]

Having the press monopolize the telegraph lines early in in the industry's existence was initially a benefit for the telegraph, since much of the industry's capacity remained idle throughout the day. However, once the public began to use the telegraph on a regular basis, it became clear that there was not enough transmission time for all purposes. The other problem was lack of optimal income for the telegraph companies by allowing the press to have more time on the lines than the public, whose messages were shorter and more expensive.

The situation was not in the best interests of the press either. Under the prevailing arrangement, a reporter could lose out to a rival who arrived at the telegraph office first. Additionally, a private interest could monopolize the lines to the exclusion of the press. The result was a very uneven flow of information from the American press. The *Rochester*

Democrat argued in 1847, "Under existing circumstances serious difficulties occur in consequence of there being but one or two wires connecting with the great cities, an individual who reaches the office first, and is anxious to transmit the contents of a long document, is able to monopolize the Telegraph."[7]

Since Morse's first come, first served did not work, it was supplanted with the fifteen-minute rule. This rule, wrote William Shanks, a journalist of the time, "required that in the transmission of the telegrams to the press, the operator should send to one paper for a quarter of an hour, than another for the same period, and so on until each was served in turn."[8] With this system, there was not much sense in trying to beat the competition because everyone was treated equally. According to Shanks, the fifteen-minute rule "virtually put a stop to news enterprise." It "placed all papers on a par; the slow ones having an equal chance with the enterprising ones."[9]

Since the fifteen-minute rule put every paper on the same level, the outcome was that the reporters began to duplicate each other's dispatches, and since this was the case, the editors decided to "take turns to receive and manifold the news." This arrangement "merged into the associated press."[10]

Getting the news first in the telegraph office was not the only problem that the press faced, there was also competition for gaining foreign news. New York Harbor was the scene of a battle for news involving the dispatching of fast newspaper owned boats that would meet ships coming in from abroad, to acquire the international news first. In an effort to gain foreign news in the same way as domestic news was now being accumulated, five New York dailies agreed to combine for the purpose of acquiring foreign news on incoming ships. They were the *Journal of Commerce*, the *Courier and Enquirer*, the *Sun*, the *Herald*, and the *Express*.

The group purchased a boat, the *Newsboy*, met ships off Sandy Hook, and collected the incoming news. At the same time, smaller boats gathered news in the harbor. With both of these efforts in place, the New York newspapers significantly expanded the scope of the pre-existing association for telegraphic news. In 1849, the *Tribune* joined the group.

By the 1850, the transmission capacity of the nation's telegraph

8. The Telegraph and the Press

industry had increased significantly compared to 1846, when the newspapers had banded together. However, instead of increasing competition among the members of the press, the cooperative arrangements held. Possibly, the reason was that the scope of the enlarged network was too much for a single daily to absorb, but another reason may have been cost. It was still possible for one newspaper in the 1850s to pay the entire expense of the AP for one year—about $30,000—but the editors, seeing the expanding volume of telegraphed information coming in, may have been looking to the future.

As a comparison between the incoming news by wire and that being sent to the newspaper by mail, the New York *Herald*, on May 12, 1846, reported President Polk's declaration of war on Mexico on its first page, followed by the words "by the Electric Telegraph." On the second page the paper printed a letter from Washington, D.C., speculating on the president's next move with that nation. On June 7, 1846, the *Herald* printed the details of the American victory at Matamoros. In the same issue, there was a news item that dealt with General Zachary Taylor's precarious position before the battle. An editorial in the *Louisville Journal* commented on the problem of mixing old and new news: "Having got by telegram the news of the overwhelming Whig victory throughout New York, we are now daily receiving by mail the papers published in that State just before the election, and we cannot help being infinitely amused at the bold confidence with which the Locofoco papers speak of their party's inevitable triumph."[11]

Changes were now coming quickly, and as *Time* magazine founder Henry Luce termed it, the public wanted "fast news." Horace Greeley testified before a British parliamentary committee in 1851 that "the quickest news is the one looked to." James D. Reid also made the same point, "The public appetite had been whetted, and the chief point of interest was the telegraphic column. There was to be found in the tersest form all the major events by which the world was agitated." Not that the "slow news" of the newspaper was entirely neglected, but once the telegraphic column was read, "the paper was laid aside for the leisure hour of home."[12]

The next step in the evolution of "instant" news vs. "old" news would be the all-telegraphic newspaper predicted by the Philadelphia *North American*: "The wires of the Telegraph will be the nerves of the press, vibrating with every impression received at the remotest extremes of the country. The events of yesterday throughout the entire land will be given, as we now give the occurrences at home, to-day. The appetite for news will be whetted into greater keenness by this supply, and those journals which give most, as the *North American* does, will fare best."[13]

With this rapidly expanding amount of news now available to any newspaper editor because of the telegraph came two problems; one, already mentioned, was the increase in costs for the service and the other was managing this constant and ever growing flood of incoming information. The only certainty was that now telegraphic news gathering was here to stay.[14]

The most important part of the telegraphic news was the commercial information. This was especially true of the incoming foreign stock market reports. Menahem Blondheim, in his book *News Over the Wires, The Telegraph and the Flow of Public information in America*, tells the story of a possible act of skullduggery that may have upended the AP's evenhanded management of foreign news.

On the afternoon of November 16, 1846, a petite woman accompanied by a black-eyed girl entered the offices of the Magnetic Telegraph Company in Boston. The woman, dressed in black, handed the clerk a message written in cipher and signed "H.C. Daniels," which she asked to have transmitted to New York. The woman insisted on remaining in the telegraph office until she was assured that her message had reached its destination.[15]

Because of her insistence on waiting until her message was transmitted, she was allowed to enter the operations room of the telegraph office until it was sent. So unusual was this that her presence attracted the attention of all of the office employees, including the president of the company, Francis O.J. Smith. As it happened, this was the day when the company was expecting the steamer *Acadia*, carrying European newspapers that would be less than a week old, to dock in Boston harbor.

This information was crucial at this time because there was wild fluctuations in the commodity markets due to crop failures on the Continent. Not only were grain prices affected but also those of cotton, because the more than Europeans had to pay for bread, the less they would have to spend for American cotton. The newspapers on the *Aca-*

8. The Telegraph and the Press

dia would determine the overall winners and losers of those who traded in grain and cotton.

The men in the telegraph office believed that the message signed H.C. Daniels had something to do with the European situation because of the woman's insistence that she be assured that the telegram was received in New York and because of the name of the recipient of the wire, Jacob Little, a Wall Street operator whose fortune came not from "solid" investment, by rather by speculation. Little was one of the earliest and most successful practitioners of market manipulation, having made his money by leveraging both short sales and short sellers. He was known as "The Great Bear of Wall Street." Once her message was received, the woman scribbled some words on a piece of paper and handed it to the black-eyed girl, who immediately left the office, returning a few minutes later with a middle-aged man who spoke to the woman and then left the office, followed shortly after by the woman herself.

About a half hour later the newspapers from the *Acadia* were brought into the telegraph office and the transmission of their contents began to be sent to New York. Shortly afterwards, however, the line between Boston and New York went down. It was discovered that the line had been cut four miles from the telegraph office. The repairs were not done until the following morning. Smith and his operators immediately began to suspect the woman in the black dress to be the cause of the problem of the telegraph interruptions.

This episode occurred in November of 1846; the severing of the telegraph wires also happened in early August after the arrival of the steamship *Great Western*, and later in the same month after the arrival of the *Caledonia*. "The wires have been cut," surmised the New York *Herald*, "by some rascally speculator" who feared that its use by the public might "upset some speculation in corn or cotton."[16]

The speculator's reason for cutting the wires was simple. President Smith of the Magnetic Telegraph Company had an arrangement with the press giving them equal access to the news on every incoming steamer and exclusive right to an uninterrupted transmission of foreign news. If the line was cut, the same news, in the hands of the speculators, could reach New York by train, giving them an opportunity to profit by using it first.

The same woman appeared in the Boston telegraph office in late

February of 1847. Again, the wires were cut after she wired Jacob Little's office in New York. This time Smith's employees managed to identify her. H.C. Daniels was found to be Helena Craig, wife of the Boston newspaper reporter Daniel H. Craig.

Daniel Craig was "a prototype of the professional manager, he was one of the most effective but least known, most colorful but least appreciated, most imaginative and most controversial of the country's great system builders."[17]

Aside from being a newspaper man, Craig was a purveyor of information, who considered beating the Associated Press as a challenge, often using carrier pigeons to send the European news from the decks of incoming steamers. With his information, he would sell foreign news to the APs' own papers or to the AP itself. Thus Craig could have been the one to have cut the Boston to New York telegraph wires; however, he would only admit to have been the person who defeated the AP, not to cutting the wires.

Smith repeatedly and publically accused Craig of cutting the wires to the point where Craig sued him for libel, and won the case. The New York *Sun* agreed with the court, believing that Smith, not Craig, was the culprit.

> It is almost unnecessary to say that we don't believe a word of it [Smith's allegations]. We know so much of that Boston telegraph, the stories manufactured to order, and the outrageous manner in which it was conducted, that we are prepared for any "statement" that will shield the line [the Magnetic Telegraph Company] from deserved reprobation. "Unknown persons" cutting the wires, indeed! The *infamous management* of *that line* stamps this latest fabrication with the mark of a well known authorship.[18,19]

With telegraph lines now being erected in Canada, Craig realized that the use of the carrier pigeons had no future and accepted an offer to work for the New York Associated Press. In this position, he had complete charge of the association's operations in the foreign news field.

The next struggle involving the telegraph took place between the East and the West as opposed to the former conflict between Boston and New York. The telegraph in the West (the Mississippi Valley and Great Lakes regions) was controlled by Henry O'Reilly. Telegraph interests in the East, most notably, those of Francis Smith, jealous of O'Reilly's success, attempted to destroy him by cutting off his connection with

8. The Telegraph and the Press

Smith's eastern lines, including foreign news. If successful, this action would have hurt O'Reilly badly. O'Reilly's weapon in the battle was propaganda and his defense was that those who controlled the telegraph in the East wanted to have a monopoly over information. The New York press, already an enemy of Smith, now had a new reason to attack him as one who wished to have a too much power. As the New York *Express* put it:

> The monopoly of intelligence is too great an element of power to be vested in a human being. The magnetic telegraph is the application of electricity, with lightning speed, to the transmission of thought.... To entrust such a power to a person, or persons, is to entrust them with the highest exercise of authority over the human race.... Here, of a sudden, has crept in upon us an uncontrolled, and illimitable Power, before which all other fade into insignificance and contempt. To allow such a power to grow up, and strengthen itself, in our apathy, is to commit it to the monopoly of light, of news, of intelligence, of speculation, of trade, of commerce, nay, of the highest and dearest interests of the human race.[20]

There was another side to the story which was brought out by Sidney Morse, the inventor's brother. If the Smith side lost, they, the newspapers, would have a monopoly over the dissemination of information themselves. As far as O'Reilly's contest with Smith is concerned, O'Reilly was successful in convincing the New York State legislature to support him by passing a law that in part required telegraph companies to transmit messages in the order in which they were received, with an exception for the press "for the transmission, for the purpose of publication of intelligence of general and public interest, out of its regular order."[21]

The war between Smith and O'Reilly brought out new products. O'Reilly secured the rights to the printing telegraph invented by Royal E. House, and more importantly the Bain electro-chemical telegraph (the forerunner of the present day fax machine), which would allow for rapid transmission of long messages. Using the Bain line, O'Reilly announced that he would construct a telegraph line between New York and Boston. Smith realized that if such a line was built, he would lose the AP business because the Bain line would be an improvement over his and because the AP was controlled by his other enemy, Craig. This eventuality meant that Smith could go out of business, since he could not exist on private business alone.

After a great many schemes and counter schemes between Smith and Craig in which neither was ultimately victorious, Smith decided to fight O'Reilly in court. He realized that the suit might take months or years before coming to trial, so he took his last shot—he informed the AP that he would not allow any messages of Craig's to go over his lines; he was, however, too late, O'Reilly had completed the Bain line between Boston and New York.

With the contest over, there were two big winners, Henry O'Reilly and Daniel Craig. O'Reilly because he overcame Smith's attempt at creating a monopoly and because he proved the effectiveness of the Bain telegraph. Craig, because of his considerable talent, became the new general agent of the AP in 1851.

In his new position, Craig went about expanding the AP, making it into the major distributor of telegraphic news nationwide. The key to accomplishing this was his control of foreign news, which he strengthened by again using his carrier pigeons. He now had the birds fly from the incoming ships to Sandy Hook, New York. From there the information travelled by telegraph to New York City.

His next step was to create a system for gathering and transmitting telegraphic news so that it could be nationwide. The plan was intended to achieve "a uniformity and concert of action ... between all of the presses connected in our proposed arrangement."

> All shall contribute, in proportion to means and relative advantages to be derived, to the expense and trouble of collecting and transmitting, from one end of the union to the other, all important news—as well as that relating to commerce as to general events—the wish being to raise the standard of telegraphic reports, both as regards the *matter* and the *manner* of the same— to make them what they ought to be—*reliable for accuracy*, and the medium through which all *really important* or decidedly interesting news shall be placed before the public, with the utmost dispatch.[22]

The plan was two pronged. Foreign news would be exclusively controlled by the New York Associated Press; domestic news was to be collected by agents that the association placed in important cities. The agents were not reporters, but rather gatherers of important news given to them by their cities' newspapers, which were forwarded by wire to New York. The reports were then consolidated and sent back to the agents, who presented them to the newspapers in their area. The number of newspapers serviced was important for three reasons. The more news-

8. The Telegraph and the Press

papers involved, the more complete the news coverage, the cheaper the cost of transmitting the consolidated news reports to them, and the greater the profit.

The only problem with the idea was the telegraph companies upon which the press association was totally dependent to receive and transmit the news. Craig, of course, knew this because of his involvement with Smith and O'Reilly, but fortunately for Craig, each of their companies, in dire financial straits owing to the battles they fought for control of the telegraphic communication between Boston and New York, merged, with Smith in control.

A telegrapher on the direct wire from Washington, D.C., in the telegraph room of the *New York Times* (Library of Congress).

The merger, instead of being a problem for the AP, turned into a benefit because Smith's new mantra was that to "disturb the New York Press [would be] both madness and folly."[23] The AP shared the same attitude. "This is the end of the war," stated the *New York Courier and Enquirer*. "We have no harsh feelings towards that officer [Smith] who has certainly shown that he possesses the quality of *perseverance* in a degree rarely manifested by anything that goes on two legs."[24] In fact, peace helped both Craig and Smith. Smith was trying to sell his company and a fight with the press would have reduced its value. Craig ultimately benefited because the association had no choice but to use Smith's line.

Since Craig was trapped into using Smith's line, he began to look for an alternative. He found it in the line that operated between Boston and New York built by Royal House, which was almost out of business. The idea that the press should operate a telegraph line was not new, but it had not been tried before. Now the AP had a choice; they could buy the House line that Craig bought an option on or stay with what they had. They chose a third course by convincing Craig to buy the House line and by giving Craig their business. As a hedge, however, whenever two telegraph companies competed on the same route, the AP would give them both business and was also willing to pay the highest price for telegraphic service as long as the telegraph companies would not send competitive news messages at a lower price.

Even with this system in place, Craig believed that the future of the AP depended upon a national integrated telegraph network. To this end, in 1853, Craig attempted to interest his editors "to buy up the direct trunk lines between Boston and New Orleans, on the seaboard, and the great cities of the West and Southwest, which could have been done for less than $1,000,000." The editors "declined to permit [him] to go on with the enterprise." "Had they assented to my wishes," Craig said later, "the Western Union Telegraph Company would have been buried in its infancy," and the telegraph monopoly "would have been as a tail to the Associated Press kite."[25]

In 1856, the AP decided to split its business not between telegraph companies in competition for AP patronage, but rather East and West. The eastern company was the American Telegraph Company owned by Craig, and in the West, the Western Union Telegraph Company. The distribution of news was now a business of national scope, rather than a competition between newspaper editors and news associations. Also by the end of 1859, the American Telegraph Company had entered into a series of agreements with any wire service that could have been a competitor. Craig now controlled all the telegraphic services in the East down to New Orleans.

In 1851, the *New York Times* joined the other six newspapers that controlled the AP. This union of seven morning papers controlled the news. It was collected solely to meet their needs without any consideration of the wants and needs of subscriber newspapers. There was little attempt to disguise the fact that the object of the association was to create and perpetuate a news monopoly. As Craig bluntly stated, "We succeeded and compelled the editors to abandon their arrangements and come into ours."[26]

The biggest problem that Craig had at this point was foreign news.

8. The Telegraph and the Press

The AP had a monopoly on it which insured the success of the organization, but all the swiftness of the telegraph could not change the fact that European news was weeks late. The telegraph had conquered the land, but not the sea. The news in any paper was a mixture of fresh domestic intelligence and stale datelines from abroad.

On August 17, 1858, the situation changed. President James Buchanan received a telegram from Britain's Queen Victoria congratulating him on the completion of the Atlantic Cable. While the cable's ability to operate was short lived—it lasted until September—it showed that transatlantic instant communication via telegraph was possible.

While Craig generally did not follow political candidates, he made an exception with Abraham Lincoln. In an unprecedented move, Craig assigned a correspondent, Henry Villard, to travel with the president-elect. Villard reported on all on Lincoln's speeches during the period between his election and his last trip from Springfield to Washington to assume the presidency. At the last minute, before the train pulled out of the station, Lincoln gave a speech in which he said goodbye to his friends. The speech caught Villard unprepared and as soon as the train left the station, he went to Lincoln and explained his predicament. Lincoln took the reporter's pad and pencil and wrote out the talk he had just given, which Villard filed at the next telegraph station.

Resourceful as Craig had been, he had no precedent on which to model the activities of his association in reporting on a war. The problem was that most of AP's agents who manned Craig's scattered outposts had been hired for their ability to use the telegraph. Even so, these correspondents, in an age when many reporters were prone to describing a battle with such phrases as "a glorious, overwhelming victory" or "a strategic withdrawal before a vastly superior enemy," kept their dispatches reasonably free of gaudy, artificial heroics. The Washington bureau chief of the AP, Lawrence Augustus Gobright, summed up the AP method:

> My business is to communicate facts; my instructions do not allow me to make any comment upon the facts which I communicate. My dispatches are sent to papers of all manner of politics, and the editors say they are able to make their own comments upon the facts which are sent them. I therefore confine myself to what I consider legitimate news. I do not act as a politician belonging to any school, but try to be truthful and impartial. My dispatches are merely dry matters of fact and detail. Some special correspondents may

write to suit the temper of their organs. Although I try to write without regard to men or politics, I do not always escape censure.[27]

As the war progressed, the nation's major newspapers, especially those in New York, began to assemble teams of correspondents to cover the fighting, but no teams were as large as those of the AP. There was trouble on the home front. Gerald Hallock, president of the AP since its inception, who also owned and directed the editorials of the *Journal of Commerce*, objected to the Lincoln administration and its prosecution of the war. With war fever at a fanatical heat, a federal grand jury stepped in with a presentment denouncing the *Journal of Commerce* as disloyal and recommending that the paper be prosecuted. Since it was obvious that the newspaper could not continue to publish under such a handicap, Hallock sold the paper in August of 1861.

By 1862, news-gathering difficulties in New York were taking an unfortunate turn. There had been a steady drain of city reporters to the various fronts, making it difficult to cover local news. The dailies began hiring women to fill in the open ranks, but it was still not enough. The answer was another news organization, Stout's Agency, a local news gathering agency that with a staff of ten covered local assignments for the AP.

In 1864, Craig had to deal with both an internal and an external problem. Henry Villard, the young reporter who interviewed Lincoln on the train to Washington, had started his own news agency in the Midwest. Closer to home, a small struggling New York newspaper, the *Daily Star*, fabricated an AP release which purported that President Lincoln had issued a surprise call for an additional 400,000 troops and had also appointed a national day of fasting and prayer for victory. The bogus report was printed in most New York papers and caused the arrest of William Prime, the president of the AP. Thereafter, a special iron stamp was made by the company and all dispatches now bore its imprint.

With the war between the North and South having ended, the new one became a newspaper war between the East and West. The problem was that the western papers were at the mercy of the AP, since they only received a minor proportion of the total available news. The western papers had banded together as the Western Associated Press, and asked the AP if they could receive a larger and better prepared report. The publisher of the *Chicago Tribune* Joseph Medill described the result:

8. The Telegraph and the Press

We succeeded in being allowed to put a news agent in the office of the New York Associated Press with authority to make up and send a three hundred word extra dispatch to afternoon papers and a one thousand word message to be put on the wire after 10 p.m. for the morning papers. It was called the midnight dispatch and was publish in an extra edition. We secured it at low tolls. The extra day dispatch was comparatively expensive as the wires were occupied at that time on commercial business.[28]

The end of the war also signaled the expansion of the telegraph and news gathering organizations to the West. The telegraph came under constant attack by Indians, who knew that it was used to summon soldiers and therefore destroyed as much of the wires and poles as they could, and the cowboys, who used the telegraph insulators for target practice. But the poles and wires kept moving West.

Also at this time there was an emphasis on an impartial gathering of the news, at least as far as the use of telegraph news is concerned. This was true for all sections of the country.[29] By the late 1860s, business and commerce as well as newspapers were increasingly dependent on the fast message transmission of the telegraph. One congressional report of the period concluded that "the business of the country is dependent on the use of the telegraph, and its suspension for a single day would bring loss and disaster greater than the entire value of the telegraph lines.[30] Historians have pointed out that newspapers after the Civil War used increasing amounts of wire news—news which presumably was relatively unbiased politically because it was sold to newspapers of different political faiths.[31]

The Western Associated Press was moving also, but its aim was to separate itself from New York without causing any problems for itself. To begin, they moved to ally themselves with the southern newspapers by inviting them to their meetings. Next, they allowed the smaller papers in their association to only pay for the news that they needed, instead of having to match the charges that wealthier papers paid for getting all the news.

While the western papers were making their moves and deciding whether to become independent of New York or join it wholeheartedly, the AP was rocked by a bombshell. Daniel Craig either left the organization or was fired. On top of that, he telegraphed all of the newspapers in the AP that he was starting his own news association. The Western group was now in a no-lose situation—either New York must grant them every concession that they demanded or they would join Craig.

After the threat from the West was made, the answer from the president of the AP was as follows:

> The New York Associated Press was founded by six publishers who have sponsored organized news gathering since 1848. We have facilities for carrying on the work and we do not propose to delegate any of our authority. It is unthinkable that an outside group should presume to feel that it can have any voice in our affairs. News gathering is our business enterprise and we do not propose to share it with others. Consideration of your plan would imply that the Western Associated Press is entitled to be treated as an equal, and that would be an intolerable humiliation.[32]

The response from the West was immediate since they were prepared for this eventuality. The Western Associated Press broke with the AP and joined Craig. And Craig had lost none of his skill, capturing the business of newspapers not only in the West, but also in AP's bailiwick New York City, including all the papers in Brooklyn and three of the nine in Manhattan. Additionally, Craig was named general agent of the Western Associated Press.

Initially in the battle, the Western Association drew ahead of its counterpart in the East, mainly due to Craig's actions in gaining new customers. But the East made the contest even by having their own correspondent in Europe, which increased the quality of the dispatches from that continent to the point where the East led in foreign news and the West in domestic.

To break the impasse, both sides met at the end of 1866 to reach an agreement favorable to both. The agreement—brokered by Western Union and signed on January 11, 1867—created a peace, but one that was favorable to the East. The AP retained its control of the Atlantic Cable, Washington news, and the New York financial news, and it was still in a position to control the news output of the regional news associations. The West gained some financial benefits and a limited degree of recognition, but in return had to dismiss Craig. Also, they did not receive a voice in the New York operation and could not receive news from other sources, and finally, there were no specifications concerning the quality of the news that they received.

By 1869, there was a new problem for the AP caused by their tele-

8. The Telegraph and the Press

graphic partner Western Union. It was reselling commercial and market news abstracted from the AP report, and although close alliance with Western Union had undeniable advantages in assuring the best communication facilities, it left the news association venerable to the whims of the telegraph company. If it wished, Western Union could selectively change any news coming across its wires. At one stage the president of Western Union acknowledged before a Congressional investigation that the New York Associated Press was under an agreement to use its wires exclusively and that all papers receiving its reports were forbidden to have any dealings with rival wire systems.

Western Union was in many respects as strong in its field as AP. They were denounced in Congress as "co-conspirators in building a press monopoly." One congressional committee claimed that by the late 1860s Western Union transmitted 90 percent of all telegrams and controlled 93 percent of the total number of miles of telegraph wire. Furthermore, Western Union, through agreements with press associations, particularly the New York Associated Press, transmitted nearly all the nation's news.[33]

News was distributed to newspapers in three main ways by Western Union. First by a press service over a regular telegraph line. Secondly, a press service could send news over a leased telegraph line. And finally, the wealthier newspapers could afford to have their correspondents send in to them press messages called "specials" over regular telegraph lines.

Because press associations (but mainly AP) had exclusive control over the amount of news carried over leased lines, Western Union records do not give any estimate of the amount of news sent this way. However, sketchy estimates of the amount of press association news and "specials" sent to newspapers over regular nonleased telegraph lines do exist for several years. In 1869 these two types of news amounted to a total of approximately 2,267,000 messages.[34] In 1880, the total of these two types of news amounted to some 2,484,000 messages, and by 1887, the total jumped to 24,667,000 messages.[35]

Two reasons apparently account for the increase in the number of press messages carried by Western Union in the 1880s. First, the increase in miles of wire from 1880 (233,534) to 1884 (430,571) was relatively larger than any other increase during a four year period from 1868 through 1900. For Western Union, this represented not only an increase

How the Telegraph Changed the World

in newly constructed facilities, but the purchase in 1881 of all the lines of the American Union and the Atlantic and Pacific Telegraph companies.[36] William Orton, president of WU, argued in the 1870s that the development of the telegraph depended more upon increased facilities than upon any reduction in rates.[37]

The second reason perhaps weakens the assertions of President Orton. It is a fact that the relative cost of telegraph news to newspapers declined during the late 1870s and early 1880s, perhaps because the increased facilities enabled Western Union to do a greater volume of business. In any event, the average tolls for all telegraph messages declined steadily from 1868 to 1900. In 1876 the average cost of a telegraphic message was about 51 cents; in 1884 it was 37 cents, and these were average costs; Western Union could and probably did charge press associations and newspapers sending "specials" a lower rate.[38]

As one example, the amount of money spent for telegraph news by the Boston *Evening Transcript* in 1879 ($8,300) represented slightly more than 5 percent of the newspaper's total costs for the year ($163,800). Wire news cost about the same percentage of the total costs in 1880. In 1881 however, the amount of money spent by the *Transcript* for wire news dropped to under 4 percent of total costs, and in 1883 to under 3 percent. This relationship then did not vary much for the *Transcript* throughout the remaining years of the nineteenth century.[39] If, as seems likely, the price for telegraph news declined similarly for the press in general, this lower price presumably worked as an inducement for newspapers to carry more wire news, much to the benefit of Western Union as the largest telegraph company.[40]

In addition to the Western Associated Press, a number of other press associations were formed during this period that also competed with the AP, which resulted in more wire news being potentially available to the newspapers. As an example, the United Press was started in 1882; it lasted until 1897. The Laffan News Bureau ran from the middle nineties through 1916. The modern United Press was started in 1907 and the International News Service in 1909 (now combined in the United Press International).[41]

The combination of Western Union and the press services was now becoming necessary. In 1879 Western Union leased its first wire to the AP. The wire linked New York, Philadelphia, Baltimore and Washington.

8. The Telegraph and the Press

AP leased an additional wire in 1884. In 1888, Western Union reported that it had leased out some 12,500 miles of wire exclusively for press reports.[42] By 1900, the AP had some 700 members, spent almost two million dollars annually, and sent approximately 50,000 words a day over thousands of miles of leased wire.[43]

All of this enormous usage was necessary. In 1872, Charles Seymour, a prominent La Crosse, Wisconsin, newspaperman, said that subscribers threatened to stop their papers unless they were filled with accounts of "earthquakes, tornadoes, conflagrations, long-tailed comets, falling meteors, explosions, shipwrecks, collisions, disasters, calamities, pestilence, murders, robberies, commotions, revolutions, wars, falling of dynasties, consolidations of empires and corporations, conflicts, strikes and rivalries, discoveries and explorations, corners, panics, creeds, platforms, philosophies and conventions, funerals, elopements, weddings, and an unceasing round of exciting, thrilling and astounding events."[44]

In the 1870s, another editor observed that sensationalism in the news had become an established fact; the public demanded it.[45] According to one editor in 1877, journalistic enterprise had come to mean the "acquisition of the greatest amount of information and the utmost celerity in laying it out before the public."[46]

In the 1870s, also, editors noted that readers eagerly demanded telegraphic news. One editor pointed out in 1873 that telegraph dispatches must be published with little regard to their news value. People read such dispatches, continued the editor, without regard to their importance because readers associated telegraph messages with messages sent only in vital emergencies. The reader appetite "grows on what it feeds on like novel reading in the young, or a whisky or an opium appetite in the old."

These trends, which were visible by the 1870s, increased reader demand for news, particularly wire news, and the demand for greater variety in content continued into the 1880s and after. To one editor in 1885, the newspaper had become "an encyclopedia, a poem, a biography, a history, a prophecy, a directory, a timetable, a romance, a cook-book, a guide, a horoscope, an art critic, a political resume, a ground plan of the civilized world, and a low priced *multum in parvo*."[47]

Finally, in addition to the demands of the readership for telegraphic news, there was also the factor of the increase in the size of newspapers brought on by decreased newsprint costs, development of the linotype and improved presses, larger audiences and more advertising support resulting from urbanization, and better distribution facilities by means of railroads. Given this situation, hard-pressed news editors undoubtedly welcomed the steadily incoming telegraphic news as they tried to fill up increasing numbers of pages.

9

Western Union

In May of 1844, the first telegraph line in the United States was erected between Washington and Baltimore. This construction was made possible because of a $30,000 investment given to Samuel Morse by Congress. Morse was convinced that the telegraph business was properly a function of the government and offered to sell the patent rights to the American people.[1] The offer was refused.

Before the offer was refused, however, Morse's partner, Alfred Vail, became involved in the possible sale. At the time, the inventors still hoped that the government would purchase and assume the control of the system, but no movement was made by Congress. At the same time, Morse and Vail received an offer of $200,000 for it. Morse gave Vail the responsibility for either selling the telegraph or keeping it under their control. Not knowing what to do, Vail sought advice from Henry Clay:

> The questions upon which I ask advice are: Whether or not the government should have the refusal; and the probability of the government's taking it. I do not pretend to be more patriotic than my neighbors, and I believe I look to my own interests closely, as every man should do, and I profess to have an innate and abiding sense of duty and care for the prosperity of my country and the perpetuity of the liberties we enjoy.... The questions are propounded and the opinion solicited for myself alone, for my own satisfaction before I act in the premises, subject to any suggestion you may be pleased to communicate.
>
> Your obedient and very humble servant,
Alfred Vail

Clay's reply to Vail states that he is in favor of a government purchase of the telegraph, but doubts that the Congress will spend the money to acquire it:

Ashland, 10 Sept., 1844
Dear Sir:

How the Telegraph Changed the World

Absence from home and pressure of a most burdensome correspondence have delayed the acknowledgement of the receipt of your favor of the 15th ultimo. I should be most happy to give you a satisfactory response to your inquiries respecting the electromagnetic telegraph, but I fear I can say nothing that will in the least benefit you. Assuming the success of your experiments, it is quite manifest that it is destined to exert great influence on the business affairs of society. In the hands of private individuals they will be able to monopolize intelligence and to perform the greatest operations in commerce and other departments of business. I think such an engine ought to be exclusively under the control of the government, but that object cannot be accomplished without an appropriation of Congress to purchase the right of the invention. With respect to the practicability of procuring such an appropriation from a body governed by such various views, both of constitutional power and expediency, you are quite as competent to judge as I am. As the session of that body is nigh at hand, I submit to you whether it would not be advisable to offer your right to it before you dispose of it to a private company or to individuals. If I understand the progress of your experiment, it has been attended with further and satisfactory demonstrations since the adjournment of Congress. I am,

                Respectfully, your friend and obedient servant,
                H. Clay[2]

During the next ten years, hundreds of small telegraph lines were established under Morse's patent rights. Morse could not attract large investors initially partially because they were skeptical about losing money on the invention, partially because they feared that the government would ultimately become involved, and partially because Morse demanded too large an interest in the venture. As a result, the principal means of raising capital for the company was by subscriptions from small investors who resided in the rural communities of New England and the underdeveloped Midwest. After this the smaller lines were gradually merged into a few large systems.

By 1857, the large systems numbered six: the American Telegraph Company, covering the Atlantic Coast states and some of the Gulf states; the Western Union Company Telegraph Co., covering the states north of the Ohio River and parts of Iowa, Kansas, Missouri, and Minnesota; the New York Albany and Buffalo Electro-Magnetic Telegraph Company, covering New York State; the Atlantic and Ohio Telegraph Company, covering Pennsylvania; the Illinois and Mississippi Telegraph Company, covering sections of Missouri, Iowa, and Illinois; and the New Orleans

9. Western Union

and Ohio Telegraph Company, covering the southern Mississippi Valley and the southwest.[3]

Ultimately one company survived in the North, the Western Union Telegraph Company, which had been chartered by the New York State Legislature in 1851 as the New York and Mississippi Valley Printing Telegraph Company with Hiram Sibley as its president. Sibley, in order to create his company, bought up rivals that had gone bankrupt, or issued new company stock to absorb rivals. The company also benefited by working with the railroads. The federal government aided the new company with a subsidy of $400,000.

During the Civil War, the value of long-distance telegraphic communication for both commercial and military use was effectively demonstrated. Western Union in particular, because of its geographic position in the North, enjoyed a great boom in business, while several of its rivals in the South were severely hampered by military operations. The company did so well, in fact, that it declared not only stock but also cash dividends.[4] Additionally, in settlement of wartime claims, Western Union acquired thousands of miles of telegraph lines built for military use given to it by the federal government.[5]

By 1866, Western Union was established as one of the first industrial monopolies, and in order to accomplish this, it acquired its two remaining rivals—the American Telegraph Company and the United States Telegraph Company—by inflating its stock and exchanging it for the stock of the other two companies. The takeover of one of these companies, the American Telegraph Company, allowed it to gain control of the first transatlantic cable line. In addition to the eastward expansion because of the Atlantic Cable, Western Union also completed a line to the Pacific Coast which began operating in October 1861. Sibley envisioned the line to California as the first step linking the European telegraph network with the United States through Russia. Western Union only abandoned the project after the successful operation of the Atlantic Cable. The California line was, however, highly profitable.

In the five years between 1861 and 1866, Western Union had evolved from regional telegraph company headquartered in upstate New York to an integrated system operating coast-to-coast from an office on Broadway in New York City. Because of these monopolistic actions, the Senate adopted a resolution asking the postmaster general for an opinion on the advisability of establishing a government run telegraph system. Since the report was unfavorable and in an effort to spur competition, Congress passed the Post Roads Act in 1866. Under the terms of the act, the telegraph companies that accepted its provisions were granted rights of way across public lands, post roads, and navigable streams. They were also granted the privilege of taking land, timber, and stone from the public domain for construction and maintenance of lines and stations; however, the government reserved the right after five years to purchase any telegraph line.[6]

Western Union along with other telegraph companies took advantage of the act, and although Western Union later claimed that it never availed itself of the land and timber privileges, it did benefit from the right to cross state lines. An even bigger benefit was the fact that it was now free from government competition.[7]

During the next thirty-five years, Western Union continued to dominate the field. Until approximately 1885 the telegraph was the only means of rapid communication. When faced with a competitor, the company, after decreasing its rates, not only kept its business, but more often than not took over the competitor, thereby increasing its size. An example of this was the United States Telegraph Company, which attempted to build a telegraph system spanning the entire nation. The company hired William Orton, head of the U.S. Internal Revenue Commission, as its president in 1865. By 1866 Orton, realizing that competing with Western Union was ruinous—the company was losing $10,000 per month—entered into a merger with Western Union. Orton's explanation, "The Co. never made a dollar—and the further its lines extended the more it lost." After the merger Orton became a vice-president of Western Union and in 1867, its president. In 1869, Orton, referring to the acquisition of its competitors in his annual report to the stockholders, stated,

> Consolidation yielded reduction in the rate for both public and private dispatches allowing for profitability.... Instead of several repetitions of messages between the great commercial centers of the country, as formally, transmission is now in most cases direct and instantaneous; and the operation of our system over the vast territory covered by our lines is fast assuming the certainty and uniformity of mechanism. Not only, however, have the public gained in time and greatly increased facilities by these consolidations, but they have received the benefit of large reductions in the rates for both public and private dispatches.[8]

9. Western Union

In 1884, Thomas Edison and Erastus Wiman, president of the Great Northwestern Telegraph Company of Canada, looked at the idea of starting a telegraph company that would focus only on providing service between the nation's sixteen largest commercial centers, but decided against the venture because even though it was the cream of the telegraph business, it was only 13 percent of Western Union's total and therefore such a company would be "a total failure."[9]

Two years after the Civil War, the company sent 6,000,000 messages; in 1893 this number had swelled to 66,500,000 messages.[10] Also according to the company's own figures during the period from 1866 to 1885, Western Union's net profit ranged from 30 percent to 40 percent.[11]

During the period from 1866 to 1899, while the company expanded both in the addition of telegraph offices and the amount of wire erected, it also lowered its average message rates from $1.09 to 30 cents.[12] However, the rate reduction was not enough to cleanse Western Union's image in the minds of the public. The criticism of the company was universal: boards of trade, chambers of commerce, state legislatures, city councils, labor unions, and newspapers of every political opinion. Because of this pressure, bills were introduced in every session of Congress between 1870 and 1896 to create a government telegraph system.[13] At least twelve times during that period congressional committees reported in favor of government participation in the telegraph business.

In 1947, H.H. Goldin compiled a list of the charges against Western Union at that time (1866 to 1899) which he had gleaned from sources such as newspapers, Congressional resolutions and petitions, and public forums. Among these charges were the following:

> High rates which had the effect of limiting telegraph to commercial uses. Whereas in European countries telegraph was extensively used by the general public for social correspondence, in the United States such use comprised less than 5 per cent. Moreover, Western Union's rate structure was replete with examples of discrimination between the various regions and between particular communities within these regions.
>
> A low quality of service, including errors in transmission, failure to maintain the secrecy of messages, and many glaring instances of long delayed messages.
>
> Granting free telegraph privileges to public officials to influence pending legislation.
>
> Utilizing access to telegraph files for speculative purposes.

Establishing contractual arrangements with the press associations and thus creating a telegraph-press combine.

Forming long-term exclusive contracts with railroads, hotels, and other agencies which placed competitors at a severe handicap.

A conspicuous failure to advance the telegraph art. At the end of the nineteenth century, Western Union's operations were still based almost entirely on the Morse key.

Burdening its employees with low wages, long hours, and unhealthy working conditions.

Stock watering and excessive profits. In 1884, when Western Union's capitalization was approximately $80,000,000, a senate committee estimated that the company's plant could be reproduced for $30,000,000. While only $2,000,000 of its common stock was issued for cash sale, Western Union between 1866 and 1899 distributed approximately $40,000,000 in stock dividends and about $57,000,000 in cash dividends.[14]

Western Union did have its supporters, however. Probably the most prominent was David Ames Wells, an American engineer, textbook author and economist. In 1873, he wrote *The Relation of the Government to the Telegraph*.

In this book Wells refuted the charges made against the company beginning with the suggestion that the government look to take over Western Union. He writes that in fact, Western Union was offered to the government twice and twice the offer was rejected. In 1844, after the government spent $30,000 to build a line to test the concept, the postmaster-general, Hon. Cave Johnson, reported "that the operation of the telegraph between Washington and Baltimore had not satisfied him that under any rate of passage that could be adopted its revenues could be made equal to its expenditure."[15] Also, he points out that more than twenty years later a proposal was made to Congress to unite the postal and telegraphic systems and the Hon. William Dennison, the postmaster-general at that time, after investigating the proposal concluded, "As a result of my investigation, under the resolution of the Senate, I am of the opinion that it will not be wise for the Government to inaugurate the proposed system of telegraph as a part of the postal service, not only because of its doubtful financial success, but also its questionable feasibility under our political system."[16]

Wells then goes on to say that these were the occasions when the federal government should have stepped in to help the companies involved and then could have benefited from their success, not now,

9. Western Union

however [1873], when individuals have built the system to its current size.

The second point that Wells mentioned was that the federal government has never become involved in the telegraph companies except for the Post Roads Acts which benefited the government because of the telegraphic connection with the Pacific Coast and with the various military stations in remote areas that the company's provided. Additionally, Congress "has hitherto steadily declined to pass any acts of incorporation, or to legislate in favor of any one Company as distinct from the rest,[17] and has thus, to all intents and purposes, distinctly recognized the policy of leaving the creation and control of telegraphic companies exclusively to the States."[18]

The next point has to do with the funding given to the railroads in relation to the monies given to the telegraph companies by the federal government:

> Any review of the experience of the telegraph in the United States would, furthermore, be imperfect, which omitted to call attention to the circumstances that, while the railroad system of the country has grown up under the stimulus of grants by Congress of money of millions of acres of its public lands, superadded to subscriptions, bounties and exemptions innumerable from States, counties and municipalities, the telegraph had given rather than received favors from the public. For if we add to the trifling donation made by Congress in the outset to Professor Morse, the highest estimated money value of all the privileges since granted by the Government to all the Telegraph Companies, namely—the right to string wires along military and post roads, to cross the wilderness of the plains, to preempt land for actual occupation—a right granted to every citizen—it is capable of direct proof that, through work performed for the Government and never paid for, and through services rendered to the public in time of war, flood, pestilence and conflagration, to the coast survey and science generally, for all of which payment was never asked or expected, the Government and the public a compensation which, to say the least, has been the nature of a fourfold equivalent.[19]

Wells goes on, covering taxes: Telegraph companies in Europe are exempt from paying them. In the United States, all telegraph companies paid from 1865 to 1871, in addition to state and local taxes, the sum of $1,549,000 in taxes to the federal government. Concerning expenses, he said: Since 1866, Western Union added more than 30,000 miles of additional line, 75,000 miles of wire, and opened over 3,000 new stations;

and regarding the benefit to the public, he observed that although virtually every employee of the company has had a salary increase since 1866, the cost of sending a message has dropped every year.

Wells' defense of American telegraph companies also included the monies paid out in dividends to its stockholders totaling approximately five million dollars between 1866 and 1872 and a comparison of the telegraphic facilities of government controlled telegraph services in Europe

Western Union telegraph building, New York, general operating department, circa 1875 (Library of Congress).

9. Western Union

and privately owned ones in the United States, showing that the latter are very far in advance of the former.

He concludes looking at the telegraphic system as a whole in the United States, "That the public, furthermore, are either not dissatisfied with the manner in which the telegraphic services of the country has been performed, or take but comparatively little interest in the subject, is sufficiently shown by the circumstance that up to the present time not a single memorial or petition has been presented to Congress from any Chamber of Commerce, Board of Trade, Municipal Organization, or Press Association which sets forth the necessity or prays for the provision of any new system, or which complains of and seeks redress for any existing grievances."[20]

Whether it was supporters like Wells, those who believed in a laissez-faire policy, or simple inertia, there was no concerted action on the part of Congress to challenge Western Union. While the company was chartered at the state level, any effort to curtail its activities was ineffective because of the interstate nature of its business. The only check to the dominance of Western Union was the introduction of the telephone. Even here, the issue was in doubt because the company acquired key patents giving some control over the telephone and even built exchanges in several major cities.[21] The situation was settled when in 1879 the Bell interests came to an agreement with Western Union that in return for relinquishing the telephone field for seventeen years, Western Union would have exclusive rights to the telegraph.[22] Western Union's main concern at this time was to prevent the Bell Company from entering the telegraph field. The agreement provided that "the principal features are carefully framed provisions for the protection of the Western Union telegraph system in telephonic patents, its telephones and telephone exchanges. This agreement also provided for Western Union to receive a certain proportion of rentals or royalties which should come to the Telephone Company. Western Union viewed this contract with considerable satisfaction."[23]

In addition to the contest with Bell, there was also a struggle for control of Western Union between William Vanderbilt, who had been the dominant power in the company since 1878, and the financer, Jay Gould, who began his drive to control the company in 1874 by acquiring a small competitor, the Atlantic and Pacific, and then used his access to railroad rights-of-way to expand it into a major competitor. When Western Union bought Atlantic and Pacific in 1877 from Gould, he gained a large block of Western Union stock. In 1881, he acquired working control of the company, replacing Vanderbilt.[24]

Within the next few years, Bell's telephone lines were completed between New York and Boston and between New York and Philadelphia, weakening Western Union's hold on rapid intercity communication. This development points out the failure of Western Union to stay completely with the telegraph. By 1900, the company was beginning to feel the competition of the telephone, an alternate means of communication that it could have controlled.

At this point, Western Union's policy of undercutting competitors by dropping prices until the competitor was forced out of business was used against them by the Postal Telegraph Company owned by John W. Mackay. Mackay was a miner who had gained a tremendous fortune in gold and silver from the Comstock Lode. The name of his company is significant; it was inspired by a plan for partial government participation in the telegraph business which had been advanced continuously since the 1860s. The plan involved offering postal space, employees, and equipment to any competitor of Western Union that promised to slash telegraph rates. Additionally, the Postal Telegraph Company entered into a rate war with Western Union in both the landline and cable business between 1885 and 1888. Western Union slashed its rate from seventy-five cents per word to twenty-five cents, which Mackay matched. Only when Western Union dropped its price to twelve and a half cents per word did Mackay give in.

Instead of selling out, however, Mackay went to Western Union's customers telling them that if their company won the rate war, it would raise its prices back to seventy-five cents. Because it was sustaining losses of three quarters of a million dollars a year, Western Union backed off. Jay Gould said of the situation that it was impossible to beat MacKay: "If he needs another million or two he goes to his silver mine and digs it up."[25] Congress turned down Mackay's plan to partner with him in the telegraph business. As a result, he continued absorbing small insolvent companies to form an extensive domestic telegraph system that by 1928 became the International Telephone and Telegraph Company.

David Hochfelder points out that at the end of the nineteenth cen-

9. Western Union

tury, Western Union's critics charged that the company routinely retarded the pace of technological progress to keep competitors out of the industry. This was especially prevalent after Jay Gould's takeover of the company in 1881. Thomas Edison was quoted on the situation, "I knew that no further progress in telegraphy was possible and I went into other lines." Henry George testified in Congress that Western Union paid electricians "*not* to invent" and that the company acquired "invention after invention simply to hold them idle."

The Postal Telegraph Company was also accused of avoiding innovations. At the 1898 meeting of the Franklin Institute, one of the oldest centers of scientific development in the country, Reginald Fessenden, a Canadian pioneer in the field of radio, remarked that it was "generally understood" among engineers that Western Union and Postal "do not want any improved apparatus" and that they only purchased patents to prevent competitors from adopting them. In 1902, Michael Pupin, a Serbian-American physicist who had created a means of extending long distance telephone communication, charged that both companies provided the public with lackluster and expensive service because they used "antiquated methods." Historians talk about the late nineteenth century as a great age of mechanization in American manufacturing. The telegraph, however, was never part of it, since it used only tiny electromagnets powered by weak current, generated in most cases by batteries.

Fessenden's and Pupin's criticisms were probably merited. Western Union had only adopted two improvements between the Civil War and the turn of the century—Edison's quadruplex[26] and the stock ticker. In contrast, some of Western Union's competitors used technological innovations. As an example, the Atlantic and Pacific Telegraph Company used a high speed automatic telegraph designed by Edison to handle press dispatches for the American Press Association.[27]

Another criticism of the company was in its use of franks, which were passes for free telegraph use. This was especially true when they were given to state and federal legislators. This was apparently done as a means of fending off hostile legislation at a relatively cheap cost. Orton directed managers to overlook free telegraphing done by important legislators such as Schuyler Colfax, chairman of the Committee on Post Offices and Post Roads in the House of Representatives, and James G. Blaine, the speaker of the House. In 1869, Orton apologized to New

Jersey Senator John Potter Stockman for "unnecessary severity" on the part of his employees in limiting his free telegraphing. In 1873, Orton told his shareholders that government officials accounted for nearly a third of the company's free business. In 1884 Western Union's vice-president, John Van Horne, testified before the Senate that the company provided about one million dollars in free service.[28]

This situation—one of supposed greed and too great a control of one of the nation's means of communication by one company, Western Union—brought forth calls for what became known as "postalization," or public ownership of the telegraph and telephone companies. Postmaster General Albert S. Burleson prepared a report and gave it to Congress on January 12, 1914. In the report, the Post Office Department compares the number of letters received per person per year worldwide with those received in the United States, and the telegrams received in the United States compared with those worldwide. It found that the United States led the world in letters received, but eight other countries, including New Zealand, surpassed the United States in telegrams sent. "From these statistics the Postmaster General infers that private ownership of telegraphs in the United States has been less efficient, as compared with other countries, than public ownership of the postal service, and that consequently the American people might expect a more efficient telegraph service through the 'postalization' of the telegraphs."[29]

The postmaster general's argument is a continuation of conclusions drawn by reformers in the previous century such as Colorado Senator Nathaniel P. Hill, who called into question the telegraph company's profits by introducing a postal telegraph bill in 1884, one provision of which called for the Senate to investigate "whether the cost of telegraphic correspondence[30] ... has been injuriously affected by large stock dividends made by the Western Union Telegraph Company."

Senator Hill's point was that if the telegraph company was controlled by the postal service, then it would not have to pay out dividends and therefore would be on the same financial basis as the mail. Also, since the postal system had a high standard of service, a nationalized telegraph system would be its equal. The American people regarded affordable and convenient mail delivery as a right and the telegraph would join it. The reformers additionally contended that the government monopoly also covered telegraphy: had the telegraph been in existence

9. Western Union

in the 1780s, the founding fathers would have treated it in the same way as they did mail delivery. Another argument from those in favor of "postalization" was cost. During the 1840s and 1850s, the mail became more a social medium than one of business because of price reductions by the federal government. This use as a social medium expanded enormously after 1851, when the price of sending a letter anywhere in the nation was reduced to three cents. The telegraph reformers wished to have the same price reduction benefits for the telegraph.[31,32]

In February 1870, the British post office began operating the nation's telegraph system. The results supported the contention of the telegraph reformers in the United States that such a takeover in the United States would be a benefit to Americans because the average cost of a telegram dropped nearly in half, the number of messages more than tripled, the number of telegraph offices also tripled, and press dispatches increased tenfold.[33] For the reformers, this overseas success was seen as a sure sign of the benefits of the foreign system over that of leaving American telegraphy in the hands of Western Union.

Unfortunately for those who supported "postalization," the British-American comparison did not work out financially. In the years between 1870 and 1895, British telegraphy lost $25 million. The major reason was that telegraphy could not take advantage of economies of scale in the same way as a postal system could. Two hundred pieces of mail could be delivered for less than double the cost of delivering one hundred pieces of mail. The delivery of a telegram required two operators, one at each end, and the wire between the two points would be monopolized during the transmission of the message. Next, a postal telegraph system would have increased labor costs, because while Western Union used low-paid messenger boys to deliver telegrams, a postal system would use higher paid adults. Furthermore, these adults would, as in England, receive civil service protection, which would reduce worker effort.[34] British telegraph engineers Henry Fischer and William Preece noted in 1877 that a typical British post office operator handled about half the message volume that a Western Union operator did, owing to civil service protection.[35]

An organization that came out strongly against postalization was the press. The primary reason was that newspapers believed that the government could not give them the same level of quality and service as Western Union. The press and Western Union each had a business to run, and to do so profitably meant that each had to provide reliable, accurate, and affordable service (Western Union to the press and the press to the public). The press was leery of what government management by the post office would mean because of its political nature. In 1866, the postal service was the largest branch of the federal government. A New York *Times* article of January 21, 1869, asserts,

> A change of this service from the Company [Western Union] or any other organization like it, to the government, must be a change for the worst.... The clerks of the telegraphic branch of the Post Office would surely partake of the spirit which characterizes nearly all American government employees. They would be indifferent, unsympathetic, generally supercilious, disposed to shirk work, and would do not more than was necessary to retain themselves in place.... Their responsibility would be to their department chiefs.... All impelling motive to satisfy the public ... would be wanting to the bureaucratic managers of the Government.[36]

The corruption of the government bureaucracy in the postal service was more than press hysteria. The practice of awarding contracts for mail delivery primarily to contractors whose politics were acceptable to the administration began as early as 1835 in the Jackson administration when Amos Kendall was postmaster general. "Together with post office employees, these contractors and their employees constituted a sizeable army whose economic interests ensured their loyalty to federal law and administration policy."[37]

Obviously, Western Union was opposed to a government takeover of its business. The method that William Orton chose avoid unification of the company with the postal service was to gain public support for its private ownership. He did this by personal appearances at public hearings, by testifying before congressional committees, and through the use of the Western Union annual reports. In each instance, he attempted to gain public support by countering all of the government's arguments for "postalization."

On the assertion that the postal service and Western Union did the same work, Orton's argument in 1870 was that it was only true in that they both communicated ideas. The huge difference was in operations, basically because postal clerks and telegraph operators were not interchangeable. Also, the comparison of the telegraph systems in the United States with those in Europe were unfair because "the telegraphs in con-

9. Western Union

tinental Europe are owned and operated by the Government, not for the purpose of revenue, but as an element of power; and in no country at the present time is the telegraph under Government control self-sustaining."

In 1874, he expressed the view that private employees are more reliable because given that Western Union was run for profit, their jobs depend on securing results, which meant meeting and satisfying the needs of the public. Orton contended that government employees have no such motivation to satisfy the public because their pay is dependent on political partisanship, influence, or favoritism. "The salaries of the officers and employees of the Post-Office are paid as promptly and at the same rates when there is an annual deficit of six million from the operations of that Department, as if the balance were on the profit side of that account." Furthermore, he continued (addressing Western Union's rate structure), the idea that the government could create uniform rates within the U.S. and charge higher rates for transmission outside the country is not a fair comparison because the greatest distances messages may travel varies considerably less in Europe as opposed to the United States. His example, "The greatest distance that telegrams can be sent in Belgium is 175 miles, while in America they can be transmitted over 5,000 miles. Additionally, the European countries to which telegraph service was being compared (specifically Belgium, France, and Switzerland) were among the most densely populated in Europe averaging 250 persons to the square mile, while in the United States the average was 10. Would it be absurd, therefore, to demand that our rates should be the same as those of Belgium?"

In his testimony before Congress in 1870, Orton took on the subject of a government takeover of Western Union directly. He began by asking the chairman of the House Committee on Appropriations why there was a problem having Western Union continue as a private company. "It seems proper that at this point I should inquire ... whether any portion of the people of the United States, have during this session of Congress, or any preceding session within your knowledge, either by personal appearance before committees, by letter addressed to members, or by petition, requested any intervention on the part of the government in this business.... No reply was made to this inquiry. Orton suggested that the only time the government should interfere with private enterprise is when it is guilty of flagrant abuses. He stated to the committee, "The only question to be considered is, those who control its affairs administer them properly and in the interest, first of the owners of the property, and second, of the public." He also pointed out that it had never been necessary in the history of the United States for the government to intervene "in any of the enterprises undertaken by the people, and in the success of which they are directly interested. Orton ended his testimony by asking the House Committee why the government believed that a Western Union takeover would be for the public good.[38,39]

An interesting aspect of the success of Western Union has to do with its structure. At the end of the Civil War, the company, having recently expanded to the West Coast, needed an apparatus so that it could control its operations. At the time there were no businesses as large and geographically spread out except the railroads, and as far as the two companies' internal structures were concerned, they were not compatible. Another organization, however, fit very well—the military. "Because the telegraph, like the railroad, was a form of enterprise so unlike the traditional small-scale ones of workshops or merchants offices, there was but one model that could bring rational structure and discipline to the new corporate giants, and that was the military one."[40]

Some examples of the similarities: Western Union referred to the physical breakdown of its various units as "divisions." In 1870, the organizational breakdown of Western Union included three divisions: Central Division, Eastern Division, and the Southern Division. The management structure was exactly duplicated in each division. This uniformity of operating procedures was completely new in American business, but standard in the military. Company directives were labeled "general orders" or "special orders." Operators in the large offices were grouped into "squads": uniformed messenger boys were called by number and placed under "sergeants." Western Union also had a newsletter; a section entitled "The Service" which listed such things as monthly transfers, appointments, dismissals, and resignations. So that its employees would know what was expected of them, the company issued a rule book detailing how the company's business was to be conducted, as well as rules governing employee behavior. Historian Edwin Gabler, who made a study of the early workings of Western Union, pointed out that because the company was growing and changing, it could not be simply

9. Western Union

a duplicate of a military framework: "What was likely at work was a kind of management dialectic: the army had things to offer those interested in corporate empire building, but telegraphy and railroading, of necessity, themselves spurred management innovation. The two fed off of, and influenced each other."[41]

Because of this military style structure, Western Union brought bureaucracy and labor-management relations into business:

> The development of logistical control by telegraph also affected the character and practice of business management in those sectors that could benefit from it. In the decades prior to the second half of the 19th century, management consisted almost exclusively of squeezing effort from generally reluctant labor. Productivity per unit of effort increased over time because of the implementation of labor-saving technical change, a process in which entrepreneurs played a role. But day-to-day enterprise management where practical, remained almost exclusively pre-occupied with the extraction of effort. The magnetic telegraph offered a new means of cutting costs and increasing total factor productivity.[42]

During the latter half of the 19th century, the bureaucracy at Western Union continuously refined their rules not only so that everyone would adhere to the same standards, but also so that the employees would know specifically what was to be done in a particular situation. The following rules governing deliveries demonstrates this management fourteen years apart:

Rules, 1870
All messages must be promptly delivered to the person addressed, or, in their absence, to their agents, clerks, or some member of their family, and a receipt taken for the same. In all cases follow strictly the directions in the message as to *place* of delivery. If, however, a message which is addressed to a person at his place of business is received after such place of business is closed, it may be delivered to him at his residence; but unless delivered to him personally, a duplicate of the message shall be delivered at his place of business the next day.

Rules, 1884
Messages are not to be left with unauthorized persons. A message must not be left with a janitor or porter of a building for delivery by him, nor be slipped under a door, nor left in a letter-box unless the addressee has *filed* with the manager a *written* request for such delivery; nor will a messenger allow any unauthorized person to know to whom a message is addressed.

Comparing these two rules, we also see an increased emphasis on formal record keeping.[43]

The use of standardized forms also became routine so that everyday occurrences could be dealt with in the same way by everyone. This combination of rules and standardized forms was an attempt to remove any possibility for the employees to act as individuals or with creativity, as well as insuring that messages arrived intact. Job performance became routine for the sake of efficiency. "In short, rationalization might be defined as the destruction or ignoring of information in order to facilitate its processing.... One example from within the bureaucracy is the development of standardized paper forms."[44]

Specialization took other forms besides formal rule books and standardized paper forms. The appearance of the auditor in the Western Union hierarchy is very important. The increasing size and complexity of the company demanded management innovations. "In some important respects the department of audit is the one on which the executive management of the telegraph has largely depended for the care of its administration. Nothing has been so safe a guide as the logic of statistics. (Dependence on statistics was seen as more scientific and more certain.) When, therefore, the business in 1866 began to assume national dimensions, the office of the auditor became one of prime importance."[45]

The Western Union Telegraph Company was also one of the earliest conglomerates. While the company did not push for new advances in telegraphy, its diversification into manufacturing was directly related to its needs for material as well as for increased profitability. While the "make" or "buy" decision did not always favor "make," it did during the initial development of the company. For Western Union, the increased control over being able to meet its own needs meant more total control of the industry. William Orton, the company president, wrote in his annual report to the stockholders in 1873 concerning the factory that the company built on Church Street in New York City:

> [The] building is admirably adapted to the manufacture of telegraph apparatus, and is in close proximity to the permanent headquarters of the Company in their new building. The factory is now capable of supplying all the apparatus required by the Company, and its capacity can be more than doubled when required. The operations of the factory for the past six months show a small profit after deducting the interest on the investment. The apparatus made at this factory greatly excels any other manufactured in this country, and it is the superior quality of the material and workmanship, rather

than the saving in the cost of the instruments, which constituted the great inducement for establishing it.[46]

At the base of the bureaucracy was the operators. In a large sense Western Union was in an "equal opportunity employer." They would hire anyone, including women and children who had the ability to send and receive messages quickly. By the 1870s the ratio of men to women at the Western Union main office in New York was two to one; however, women were regarded as "admirable manipulators of instruments" well suited to telegraphy (since it wasn't too strenuous), and they could spend the quiet periods reading or knitting. The hours were long. Most operators, including the women, worked ten hours a day, six days a week. In many cases, female operators were segregated from the men, and sometimes a "matron" was employed to keep an eye on them.

But while the female operators were physically isolated from their male counterparts, they were of course in direct contact with them over the telegraph network throughout the working day. As a result many working relationships flowered into on-line romances. According to one writer, "Sometimes they flourished, sometimes they came to an abrupt halt when the operators met for the first time."[47] Women operators were habitually assigned to the easier wires—that is the single circuit, light traffic lines which precluded them from being able to pick up their skills by learning to operate the more complex multi-circuit lines, which would have made them, according to Western Union's organizational hierarchy, first class operators. Consequently, even when women did become operators, they had inferior skills by design, inasmuch as they were confined to these positions. The company justified this practice because they viewed women as temporary workers (until they married), hence there was no need to promote them or put them on a career track. This situation as had economic benefits for Western Union, since the least skilled positions received less pay—so the women in these positions received less pay even though they worked the same number of hours. This practice was also justified via ideology—women were not supporting families therefore they did not need a "family wage."

The question then, was, why would women take these jobs? Nineteenth century author W.J. Johnson, in his book *Lightning Flashes and Electric Dashes*, tells of a contemporary writer, W.J. Foster, who addressed this issue. His conclusion: [Telegraphy] "gives employment to women as well as men, and this assists in the practical solution of the difficult question: What must society do with the capable and intelligent female population who cannot marry, for the very sufficient reason, among others, that there are not enough men to mate every one of them?"[48] Demography also played a role, since there was a large pool unmarried women available from European immigration and among the widows of Civil War casualties.

A consequence of the control that Western Union had over its employees was a slow erosion of their telegraphers' status. During the Civil War the telegraphers organized an association for their mutual benefit. This was not a union but rather an organization whose aim was to improve professional standards, and to provide for members in the event of death, retirement, or sickness. This group, the National Telegraphic Union, avoided taking a stand on controversial issues such as the admission of women as members or the right to strike to obtain higher pay and better working conditions; also, the organization existed because the employees began to think as "national" as the company:

> National is a key word here, for the war (and subsequent reconstruction) had a marked centralizing influence on the country. Three national labor unions had emerged in the 1850s, but thirty one appeared during the 1860s and 1870s. Sectional conflict, the reintegration of a chastised South into the Union, and an increasingly powerful and activist federal government did much to make labor leaders think nationally. So did the shifting economic emphasis from local to national markets. And no industry better represented this crucial change than the telegraph, no firm better than Western Union.[49]

Postwar telegraphers earned less pay and had less control than did their predecessors before and during the war. The company cut the operators' wages from 25 to 40 percent while expecting them to handle two to three times as many telegrams per shift. Because of this situation, the telegraphers abandoned the National Telegraphic Union and began in 1868 the Telegrapher's Protective League, which agitated for pay increases and better working conditions for its members. Its primary goal was to organize the telegraph operators who worked for Western Union, which by then was the largest telegraph company. The league called a strike in January 1870 after Western Union attempted to cut the wages of four operators in San Francisco. The strike was unsuccessful, lasting only two weeks, because the company simply replaced the strikers

9. Western Union

with non-union operators. The telegraphers attempted to strike again in July 1883 under the auspices of the Brotherhood of Telegraphers, a union affiliated with the Knights of Labor. The issues were an eight hour day, wage increases, and equal pay for women. Approximately 8000 operators comprising about one-third of all the telegraphers in the nation joined the strike. This strike was called off after a month, failing to meet its objectives.

With the Civil War over and America's beginning entry into Mark Twain's Gilded Age, these Western Union employees who were attempting to better their positions at the company were part of the economic shift from agriculture to one based on manufacture, with the center of work moving from the home to the workplace. "The old Puritan ethic which stressed morality, hard work and common welfare was supplemented by the ethic of laissez-faire economics, emphasized individualism, success and competition."[50]

This new structure offered new employment positions and opportunities at Western Union. The company jobs were above blue collar and manual work and below entrepreneurial work. The telegraphers "were among the very first mass of white collar employees." Also during the transition period of the nineteenth century, the "old" middle class was giving way to a "new" middle class. The telegraphers were among the lower echelons of this new middle class.[51] The fact that telegraphy was a national phenomenon with the standardized work practices and salaries put into place initially by Western Union made the occupation an important vehicle for the emergence of this new middle class. Telegraphy was an industry large enough in both scale and scope to foster a common national identity and culture among its operators. Although not effective until the twentieth century, much of this culture was disseminated by union membership. Journals and other professional associations also helped to create and foster a common identity among telegraph operators.

In addition to the debate between a company and its employees after the Civil War, there was the struggle for power between the government and big business, with the "public" merely a pawn in the battle. The public's relationship to both business and the government changed as society underwent modernization, made possible by advances in printing as well as telegraphy, which made the public better informed.

Because of this, both the government and business took great pains to win its approval. During the nineteenth century, it was business that bested government in these contests. However, one must be careful not to assume that this was an expression of "public will," that is of the public influencing their congressmen. Rather, as Lester Lindley points out, Western Union used the lobbying of congressional members to "buttress their freedom from national restraints."[52]

10

The Telegraph Workers and Unionism

By 1863, a first class telegraph operator could earn between $110 and $118 per month. In 1870, the wages ranged between $90 and $120 per month. The variations were generally due to the amount of inflation during and after the Civil War.

Not much is known about the type of people drawn into the telegraphic occupation at the time, except that they were considered to be of higher than average intelligence. They were also better educated than the average worker during the Civil War period. For that reason, the occupation was considered more of a profession than a trade. A typical operator was described by a newspaper of the time as a "taciturn, suspicious looking individual, gifted with great clearness of head, and a precision in judgment in examining faces, far above the average.[1] The newspaper did not define what it meant by "suspicious."

At this time, the largest telegraph company in the United States was Western Union, a company that had reached its present size basically through acquiring other telegraph companies. Previous to 1866, there was no cause for worker problems in the industry. B.F. Shrimpton, a long time Western Union employee, testifying before the Senate Committee on Industrial Relations in 1914, detailed the changes in the telegraphic industry after Western Union became a dominant player: "During the first twenty years of our telegraphic development, or until Western Union obtained a monopoly in 1866, the operators had very little cause for complaint.... There were no grievances and salaries were satisfactory.... Annual vacations with pay were granted; time lost on account of sickness was not deducted; hours were reasonable and working conditions very good." As soon as Western Union took over salaries were reduced to $65 to $115 per month, vacations were abolished, lost time deducted, hours lengthened and "odious discipline inaugurated."[2]

During the nineteenth century, the strength of the operators' union efforts matched the growing strength of Western Union. Operator union efforts began during a telegraphers' meeting on October 26, 1863. While this was not an actual union organizational meeting, it was an effort to join together to assess the operators' working conditions. During another meeting on November 5, 1863, the operators formed the National Telegraphic Union. The second convention on September 5, 1864, showed the member concern over the actions of Western Union when their official organ, *The Telegrapher*, carried the following message to the members:

> We sincerely regret this consideration. Although it is money in the stockholders pockets, it (the consolidations of smaller telegraph companies with Western Union) will be the death to telegraphers generally, and an injury to the telegraph public. With the power which this great monopoly puts into the hands of its officers, salaries will decrease as surely and as naturally as water runs downhill, and the odds are so much against us that it will be useless to remonstrate. This monopoly is so vast in its proportions that no opposition can face it without enormous capital, larger than most new organizations can boast and the indulgent public must early conclude to put up with vexatious delays. Sic transit gloria opus.[3]

When first organized, the National Telegraphic Union was basically a benevolent organization concerned with such items as sick benefits and money for funeral expenses; however, conditions in the industry became steadily worse for the operators. Some delegates to the union's conventions were fired and in most cases, salaries were cut. Because of these moves the operators began to realize that they needed a protective rather than a benevolent union. The telegraph companies were not blind to what was transpiring and in 1867 launched a journal similar in name and content to the union's, which was then taken over by Western Union. Also in 1867, the National Telegraphic Union set up an insurance department, only to be followed several days later by Western Union's announcement of the inauguration of a like department.

By the end of 1868, it had become obvious that the National Telegraphic Union was not strong enough to successfully deal with a company the size of Western Union. It went out of existence in 1869. With Western Union still acquiring new companies and lowering the opera-

10. The Telegraph Workers and Unionism

tors' salaries, National Telegraphic Union members met and formed a new organization on September 23, 1868. This new group, the Telegraphers Protective League, had a preamble to its constitution modeled after that of the Brotherhood of Locomotive Engineers, stating its purpose:

> We seek to build up as a protection against the aggression of this powerful accumulation of capital, an organization of labor, founded upon well-defined principles, guided by well-directed counsels, and governed by intelligent legislation, which shall become equally powerful and equally worthy of respect. We seek not only to protect our own rights, but those of the public, by retaining in our ranks the most worthy and skillful operators in the country, instead of allowing them to be driven, through an unjust exercise of superior power wielded by willing agents of our employers, who are ever seeking official commendation by exhibition of parsimonious economy.[4]

The members were required to take an oath to "increase the power and efficiency of the League, and also a pledge to obey the rules of the organization." To show that the league meant business, it also adopted the following statement: "That we have been forced to assume an attitude of defense is our misfortune, not our fault; but now that we have undertaken the work, let it be carried out earnestly, faithfully, and vigorously, for we have rights which others must be *taugh*t to respect."

Matters were now coming to a head for both the league and the company. By this time Western Union had a firm hold on the industry, the nation's economy was not good, the cost of living was exceedingly high due to currency inflation, and the league was rapidly gaining strength through membership recruitment when Western Union decided to revise the operators' salaries.

In 1870 the combination of these factors coupled with the actions of Western Union in its San Francisco office started the battle. From all accounts, the following is the most accurate reconstruction of what transpired.

Two men who had been recently hired by Western Union had their wages cut by the amount of $5 per month, while another employee received an increase of $20 per month. One of men who had his salary decreased was the secretary of the Telegraphers Protective League, who refused to accept the salary reduction and took the case to the league. The league suggested that the matter be arbitrated, an action that the company accepted at first and then turned around and fired the two men. Management then announced that the two men had been fired "for attempting to create dissatisfaction within the office, "and that "no reductions in salaries have been made or contemplated, the only changes having been by way of increase."[5] Obviously this was a maneuver to cover up the salary reductions that had originally started the controversy.

The league demanded that the two men be reinstated under threat of strike. The demand was refused, the company stating the operators had "surrendered their liberties and judgments to become fractional instruments of an irresponsible body." The league called for a general strike on January 3, 1870. By January 5, the strike had spread to New York, Atlanta, Washington, D.C., Boston, Philadelphia, Pittsburgh, Cincinnati, Chicago, Detroit, and Portland.

The strike caused a great deal of public interest and was given a good "play" by the newspapers in the large cities. The New York *Times* and The New York *Tribune* both condemned the actions of the strikers, while the New York *Herald* sided with the workers. The *Herald* attacked Western Union as a monopoly that had disregarded the interests of the public and demanded that Congress act immediately to curb its activities. Further, it asked Congress to purchase the telegraph companies or create one run by the Post Office Department in competition with Western Union.

According to the newspapers, operators had struck in thirty-four of the principal cities in the country. In these cities almost all of the operators had left their instruments. Moving quickly, the company began to replace the striking workers by moving operators to the affected areas and bringing in retired workers and those that had gone into another line of business. This proved to be the most effective way to combat the strike. Western Union gave these workers large bonuses and in some instances double time wages.

The strikers received support from the bricklayers' and printers' organizations, while the Workingman's Union contributed $500 to support the walkout. One major newspaper, the New York *World*, initially supported the workers, stating that the strike had been called because salaries in San Francisco had been reduced and the other operators struck in support and to avoid a similar fate. On January 8, however, the paper claimed that the strikes would lead to the application of new inventions that would replace them.

10. The Telegraph Workers and Unionism

By January 12, the workers that the company had brought in to replace the strikers began to turn the tide in Western Union's favor. The New York *Times* stated editorially, "The Western Union has the situation well in hand. The union states that the company intends to reduce wages, which the company denies; the company should know what it is talking about. It seems to us that the outside organization arrogates to itself a power of decision in the premises which is not based on any substantial authority." At the same time, the other major newspapers nationwide, sensing that the strike was over, stopped covering it. On January 18, R.W. Pope, grand chief operator of the Telegraphers Protective League, issued the following statement: "To All Circuits: I have canvassed the situation, and believe it is useless to longer continue the strike. I hereby absolve all Western Union members from their oaths and advise them to return to work."[6]

There were a number of factors involved in the losing effort by the Telegraphers Protective League. Basically, unionism was too new in the United States and therefore the league made fundamental mistakes. First, the league was too loosely knit and without a plan of operation to meet a specific threat. Secondly, it had not been in existence long enough for strong bonds to have formed among the members which would have allowed them to withstand the company's pressure. Third, again because of the short time that the league was in existence, there weren't adequate financial resources to allow for a long strike. And finally, the strike came at a time when business in the telegraph industry was at a low point; as an example, there was a lack of political news because Congress was not in session.

Because of Western Union's financial strength, it was able to pay extremely high wages to replacement operators, including those employed by the railroads. This was a double blow. Not only did the league receive no support from these workers, but they were a large part of the "scabs" that replaced them.[7]

While most of the operators were given their old jobs back, they paid a high price. According to testimony before the Senate Committee on Education and Labor in 1883, Western Union, despite making huge profits, reduced worker salaries between 25 and 40 percent.

The end of the strike in 1870 was also the end of telegrapher union activity until 1882. During the intervening period of time, the idea of unionism in the United States had become more and more important because of the gradual shift from agricultural labor to that of employee labor—mainly in factories, but also in other industries such as telegraphy and railroading. This led to the establishment of the first important union, the Knights of Labor, and was spurred on by the great railroad strike in 1877.

In 1882, the telegraphers organized again, and this time attempted to avoid the mistakes of 1870. First, this new organization, the Brotherhood of Telegraphers, allied itself with the Knights of Labor, and then grouped into 150 locals nationwide with 10,000 to 11,000 members total. Next they strengthened their union by including the railroad telegraphers. In order to prevent job competition, the brotherhood constitution forbade members to "instruct any person in the art of telegraphy until the local Assembly of which he is a member shall first have granted permission to do so, and until said permission has been ratified by the executive board of the District Assembly."[8]

Unfortunately for the brotherhood, their preparations were not in place long enough before another incident caused a problem. In this instance, the problem concerned a linesman, not an operator, who was a member.

The linesman, William Sullivan, formerly a carpenter with Western Union in New York, became a linesman earning $75 per month, $5 to $10 more than other linesmen, a carryover from his work as a carpenter. He was requested to do carpentry work by the company, but because the brotherhood's constitution forbade a member to perform work other than his own, he refused. When the company told him that his salary would therefore be reduced, he resigned. A new man was hired, but the foreman who was a member of the brotherhood refused to train the new employee and was therefore fired. All of the linesmen went on strike because of this action and could not be replaced because all the linesmen in the city were members of the brotherhood, creating a tension-filled atmosphere between the union members and Western Union.

In spite of this situation, on July 16, 1883, the brotherhood submitted what was known at the time as a "memorial" to the company. This memorial asked Western Union for an eight hour working day, equal pay for women employees, and a 15 percent wage increase for all employees.

10. The Telegraph Workers and Unionism

The company announced to the newspapers that the demands of the union were "laughingly illogical," that a 15 percent wage increase was "preposterous," and that the telegrapher's wage scales averaging $75 per month for men and $45 per month for women were "equitable and just." More importantly, the company refused to recognize the right of the union to bargain for its employees.[9]

As was the case in the previous strike, the newspapers had opposite opinions. The New York *Herald* believed that the operators had acted in good faith and presented their needs in a correct manner; the problem was that they asked for too much. Also, that the operators had a right to strike and without this right, they would effectively be slaves.

On the other hand, the New York Daily *Tribune* opposed the strike. "It is difficult to see any reason for the operators striking, except that they want summer vacations. There is nothing in the condition of the business to warrant any advance at the present time. The Telegraphers Brotherhood does not embrace all, nor even a majority of the operators. Public opinion most certainly will be against a strike."[10]

The brotherhood waited three days for an answer to its "memorial." When none was forthcoming, the workers left their posts and the strike was on. As the day wore on, reports kept coming in to the New York headquarters of Western Union about the progress of the strike. By the end of the day, according to General Eckert, the Western Union general manager, two-thirds of the operators had struck.

The greatest single grievance was undoubtedly that of salaries. The operators were fully aware of the stock watering engaged in by the telegraph companies and the huge profits that they had been piling up, even during the depression years of 1881 through 1883. Of the $48,000,000 profit that Western Union had earned in thirteen years, *$18,860,286 had been made in the three years prior to the strike*. At the same time, a systematic campaign of wage reductions had been under way; according to Superintendent Humstone[11] a general wage reduction of 5 percent had been imposed in 1877 (Western Union profits during that year were $3,140127.67) in addition to individual salary retrenchments all along the line.[12]

According to testimony before the Senate committee by a number of operators, their present salaries were below the amounts that they had received in the past. The brotherhood estimated that the average pay for all operators was $54.43 per month. The average actually compared favorably with other trades; only the bricklayers, carpenters, and stonecutters received more. Another issue was the number of messages that an operator had to handle on a daily basis. Virtually every operator testified that in addition to a cut in wages, there was also an increase in messages that had to be handled.

This strike, as opposed to the last one, had immediate and widespread effects, especially on the financial exchanges. On the Maritime Exchange all ships news had been cut off; the Cotton Exchange almost closed completely. The Produce Exchange was so badly hurt that one of its leading brokers stated, "Are we in sympathy with the strikers? Yes Sir. Emphatically!"[13] There were quotations on private wires, but they were considered suspect, and were therefore ignored.

Western Union fought back as best it could under the circumstances. Every employee who knew anything about telegraphy was pressed into service. The company posted notices that it would accept all messages for transmission, subject to delay. One of the great disadvantages that plagued the company was the fact that when the linesmen went out, they took the diagrams of the wire setups with them, so it became extremely difficult to locate the source of any trouble.

As the strike wore on, the New York press with the exception of the *Herald* became solidly in opposition to the union. The strikers, however, received some unexpected support. Thomas Edison addressed the strikers offering encouragement and $300, and the Produce Exchange passed a resolution asking for subscriptions to help the strikers. The strikers themselves gained some additional support by comporting themselves in an orderly manner and refusing to take any drastic action such as cutting the telegraph wires.

On July 26, the American Rapid Telegraph Company announced that it had come to terms with the strikers, who would receive a 15 percent raise, but the news was not all good. Some western railroads were transferring their operators to Western Union in an attempt to break the strike. The union stopped this practice by stating that if it continued, they would take steps to call the railroad operators out. The press outside of New York City basically agreed with the New York papers in siding with the companies, including the Chicago *Tribune* and the St. Louis *Post-Dispatch*. The *Scientific American* however sided with the union:

10. The Telegraph Workers and Unionism

"It would be difficult to find a more able, intelligent, and industrious body of people in the world than those telegraphers. The quietude with which they have conducted their strike, and the unanimity of their ideas in respect to their demands, afford grounds for the inference that they know what they are about. The public will rejoice to see their wages increased and their hours of labor reduced, even if the prices for sending telegraph messages are slightly increased."[14]

Various religious figures also weighed in with opinions. The Rev. J.E. Searles of the Wheeler Street Methodist Church held that the union and the strikers were wrong and that great corporations were agencies of modern civilization. The Rev. Joseph Pullman of the Fleet Street M.E. Church sided with the strikers, accusing Western Union of being a slave owner and saying that the company had been robbing the people by watering its stock.

On August 1, the union suffered a serious blow when the operators employed by the American Rapid Telegraph Company who had returned to work received the same pay as before, not the 15 percent raise they had expected. Apparently, this "settlement" had been arranged by the union to put pressure on Western Union, since the union made no serious effort to explain the situation to their members.

The next problem concerned the railroad telegraph operators. The union gambled that their joining the strike could be held in reserve until needed. When they were asked to join on August 8, it was too late; the railroad operators believed that if they joined at this late date, and the strike was lost, that they would never work on the railroad again. Also, with the strike going badly, the operators were resorting to cutting wires and pelting Canadian scabs with rocks.

On August 17, the membership was told that to continue was hopeless and were advised to return to work. For the second time in fourteen years, Western Union workers had struck again their company and for the second time they were soundly beaten.

As with the first strike, the union had not been in existence long enough to accumulate enough financial resources for a long walkout. Members did not have enough experience to know what moves to make and when, as in holding back on calling out the railroad operators, and there was a lack of solidarity not only among the present operators, but also among the ones that Western Union called in, as they did in the first strike. There was also a large number of operators known as "boomers" who moved from town to town, did not belong to the union, and were glad to work for a bonus or any opportunity for extra pay. They took many of the vacated positions. And finally, the union underestimated just how tough Western Union was and failed to see they would not give in easily.

Intrinsically, the operators believed themselves to be gentlemen—professionals to whom it was unbecoming to engage in a common brawl, even though that attitude was needed to defeat Western Union.

The strike also illuminated the connection between the Brotherhood of Telegraphers and the Knights of Labor. Again, the brotherhood's biggest problem was a lack of funding, which was not relieved by the Knights of Labor, who gave them nothing. Additionally, the New York *Tribune* quoted the *Journal of United Labor*, the official organ of the Knights, condemning strikes because they were useless and ill-advised and, further, that there was no reason for the knights to give the brotherhood anything because the brotherhood had only contributed two dollars to the knights. John R. Mitchell, one of the leaders of the brotherhood, said of the situation, "We are not satisfied with the course of the Knights of Labor, and one of the results of the strike may be a severance of connection between the Brotherhood and the Knights of Labor."[15]

The possible severing of connections between the two organizations never happened. The end of the strike was also the end of the brotherhood. Also, the knights began to lose influence and were replaced later by the American Federation of Labor.

The operators lost between $300,000 and $400,000 in wages during the month-long strike, and approximately 150 operators lost their jobs or decided to find work elsewhere. Western Union claimed to have lost $276,000 in revenue and an additional $433,000 was spent on bonuses and to feed and house non-striking workers. According to the postmaster general's report of 1890, Norvin Green, the president of Western Union, believed the money to be well spent: "As I told General Eckert this afternoon, the several hundreds of thousands of dollars which have been lost in this strike I regard to be the best financial investment ever made by this company. Hereafter, General Eckert tells me, he will be able to get one-third more work out of a man for a day's services and the economy of such a step will retrieve the loss in less than six months."[16]

10. The Telegraph Workers and Unionism

While the years between 1883 and 1907 were mainly peaceful in the commercial telegraph industry, there were still underlying occasions of friction between the companies and the operators. In 1884, Western Union operators in Galveston, Texas, had their wages cut and their hours increased. The operators walked out and operators were brought in from other cities. When the *Telegraphers Advocate* exposed the test of the worker's reaction to this ploy, the operators were reinstated. During the same period, the Order of Railroad Telegraphers was organized over the opposition of the railroads, indicating that some animosity between the railroad operators and the railroads existed.

In 1887, a new telegraphers' organization was formed called the Triplicate Brotherhood. The name may have come from the copying of telegrams in triplicate. This union was replaced several years later by the Order of Commercial Telegraphers.

The *Electrical World*, in its September 8, 1891, issue, quoted Assistant Superintendent William Lloyd of Western Union on the subject of strikes. His reply, "Our men have no occasion for organizing. They get all they want and as far as is known, have no grievance.... They work seven hours per day and receive 44 cents per hour for extra time."[17] This was not exactly the case. There was a glut of skilled operators and therefore unionization was extremely difficult. For that reason another strike was never considered. "Better than a strike would be some effective discouragement of the swarms of beginners who are cajoled into learning telegraphy by lying stories."[18]

Salaries by 1890 had sunk to their lowest levels. Men were paid $40 to $45 per month and women $12 to $14 per month. A witness before the United States Industrial Commission testified in 1914 that the telegraph company believed that it was wrong for two operators to get together in order to negotiate for better wages, but it was "wrong for two or more telegraph companies to squeeze the employees and the public both."[19]

On March 4, 1895, sixty telegraphers met in Philadelphia to make another attempt to unionize commercial telegraphers. In this case, the main objective was to have the federal government take over the telegraph systems. The New York *Herald* stated that similar meetings were taking place in other large cities, including Chicago and Buffalo. On the same day, the New York *Daily Tribune* carried a story that a telegraphers' union for railroad and press operators similar to the American Railway union had been organized.

While the operators were continuing their efforts to form union organizations that would put them on an even basis with the telegraph companies, they were receiving some possible assistance from the government. Newspapers such as the New York *Herald*, Boston *Globe*, the Philadelphia *Times*, and the Chicago *Tribune*, along with a number of union and chambers of commerce, asked Congress about the monopolies in the telegraph industry. The testimony of these entities was done under the direction of Frank Parsons.[20] Parsons testified that Western Union rendered service only where it was profitable, leaving the remaining areas without telegraphic service. He also pointed out the tremendous worth of Western Union vs. the small salaries of the operators. While a number of bills were reported out of both the House and Senate to deal with this situation, no laws were passed to redress the inequities.

With no assistance from the government and a possible opening in its battle with Western Union because of the company's ongoing warfare with competing telegraph businesses, the operators tried again to form a viable union. On July 19, 1903, the twentieth anniversary of the beginning of the great strike of 1883, the operators met and created the Commercial Telegraphers Union of America. The new organization began with 60 locals and 8,010 members and was recognized by the American Federation of Labor.

The first action of the union was to ask the courts for an injunction to prevent Western Union from discriminating again its members. H.A. Estabrook, counsel for Western Union, commented on the application: "I have no fear this injunction will be granted, for a company certainly has a right to employ whom it pleases and discharge men who do not suit it. This corporation has never discharged a man without cause."[21]

The court's decision went against the company. It held that without contract relations, the company had the right to discharge, with or without notice, at any time. The same right to quit was accorded the employees. The presiding judge, John H. Rogers, said: "There is no such thing in law as conspiring to do a lawful thing." Commenting on the blacklisting of workers, he continued: "Are they [the company] doing acts that are unlawful? If so, what are they?"[22]

From this point until 1907, there ensued a kind of "cold war"

10. The Telegraph Workers and Unionism

Messenger boy working for Mackay Telegraph Company in Waco, Texas, in 1913. He said he was 15 years old (Library of Congress).

between the operators and the companies with incidents of friction and alleged discrimination. Western Union moved to meet the union maneuvers with an intensified effort to hire and train new operators. The February 1904 issue of the *Commercial Telegraphers Journal* urged all members to refuse to cooperate with the company in these efforts. In December of the same year, the *Journal* stated its immediate objectives: equal wages for equal work for both sexes, abolishment and modification of the bonus system, and better conditions for branch office operators and managers.

By 1906, the combination of a decline in business activity throughout the nation, continuous pressure from Western Union, and local strikes in Nashville and Houston began to tell on the operators.

When the break came, it was in the form of a wildcat strike. First, the operators went out in Los Angeles, and then Chicago. By August 13, the Chicago *American* estimated that 15,000 operators were on strike nationwide. Samuel Gompers, the American Federation of Labor president, issued the following statement on the situation:

> Smarting under the galling yoke of oppression, cunningly and stealthily placed upon the necks of the telegraphers of the country, by the petty tyrants and mercenary hirelings of the commercial telegraph companies, during the past twenty five years, the workers at the key have refused longer to "bend the knees" to subservient masters who scorn to recognize them while sailing southern seas in their private seraglios and sending fatuous cablegrams to workers who have spent their lives in toil that millions might be rolled up in fortunes for the idle sons of this unproducing class.[23]

On August 15, the strike was officially called, but once again the problem was finances. The union did not have the money available for a long strike against the cash rich telegraph companies. For the third time the operators had gone out unprepared and paid for it. Desertions began almost immediately and the strike officially ended on November 9.

Although the strike disrupted telegraph communications across the United States and Canada, Western Union was able to function using operators who deserted the union and scabs. The 89 day strike, according to the *Boston Globe*, cost the company approximately $500,000, but was still able to declare a dividend of 1¼ percent.

After this strike, there was another in 1919 with the same result and then a long period when the Commercial Telegraphers Union of America went into decline. In every situation, well into the future, Western Union came first. Consider the following from the President's Communications Policy Board in 1949: "A look at these figures (Western Union's balance statements), should convince the most skeptical that the company has a hard row to hoe in the next few years. Should business conditions take a downturn, the union will be faced with a challenge if the company decides to repeat its policies of the thirties, when employee wages were cut to the bone and thousands were furloughed in order to maintain uninterrupted profits during the depression years."[24]

Conclusion

The telegraph offered unprecedented capabilities to its users, as remarkable for its time as new inventions in recent years have been in conveying information. It allowed its users to send nearly instantaneous transmission of complex messages over enormous distances, freed information from dependence on human carriers, and employed a medium (electricity) that defied human understanding. Even more than the steam engine, transoceanic telegraphy created a liberation from the influence of weather conditions that imposed so much uncertainty on communication by sailing ship.

While the Atlantic cable had gradual consequences for diplomacy, its importance cannot be minimized. Telegraphy tended to increase foreign policy centralization and to integrate the international system more tightly. It also posed serious hazards for foreign ministries, most notably by increasing the occurrence of security breaches and garbled messages. Since this was the case, in spite of the telegraph's benefits, ship-carried diplomatic pouches continued to play a predominate role in American foreign policy until World War I.

This volume points out the main areas that were affected by the telegraph, but as time passed, the technology permeated other areas. In an 1869 court case, the judge allowed a man to send an affidavit by telegram. After that, it was a question of sending the accused to another city to stand trial. The judge, the Honorable J.D. Caton, came up with an idea to save moving the proceeding: "Soon it occurred to me that if I could arrest a man by telegraph, I might try him by telegraph"—which he immediately did. Testimony was given via telegraph, with the written telegram acceptable as legal testimony. Judge Caton wrote later: "That this was the first, if not the only trial, I had ever heard of being conducted by telegraph eighty miles off.... I am one of those who believe that the law is capable of adapting itself to all the improvements of advancing civilization."[1]

How the Telegraph Changed the World

It is clear that Samuel Morse never conceived of this use of the telegraph, but it is an excellent example of how people manipulate technology to their advantage. In 1873, an article in *Harpers New Month Magazine* stated, "Almost every new season brings forward some new application of the telegraph. A complete network of wires now connects the stock exchange with the broker's offices and the leading hotels, and the business done at the board is printed off in more than a thousand places in New York simultaneously."[2]

Prior to the invention of the telegraph, time in the United States was not standardized. While the railroad and mechanization began to contribute to this standardization, it was the telegraph that finally had the effect of standardizing time, making it subject to more precise management and turning it into a commodity controlled by experts and offered for sale.

This was accomplished by a cooperative effort between the federal government and the Western Union Telegraph Company. The arrangement worked as follows: Each day experts at the United States Naval Observatories in Washington, D.C., and Mare Island, California, would determine through astronomical observations the correct time. Western Union's transmitting clocks would then send out the correct time over the companies' entire telegraph system every day between 11:55 a.m. and noon. This became the standard time of the United States and was used in the operation of every railway system in the nation. While Western Union received no compensation for this service, it gained revenue from the sale of self-winding clocks, and by communicating the time to those who owned these clocks.[3]

The tempo of American life was changed by telegraphy. Instantaneous communication meant that there was less time for decision making. Responses had to more immediate. Americans, because of telegraphy, now became aware of time. New phrases entered the American lexicon such as "saving time," "having no time," "running out of time," "and being "up with the times." For example, Western Union published a pamphlet entitled "Suggestions for Social Messages via Postal Telegraph." The pamphlet's purpose, according to the company, "is merely to assist you ... to save your time and effort when it comes to phrasing your words for suitability in telegrams." On running out of time, the pamphlet suggests that a social telegram is the more expedient in cases

Conclusion

where "you may forget or postpone till the last minute to take care of these obligations." On having no time, an actual telegram from the Western Union archives dated December 24, 1880, read "All send Merry Christmas to you all. Too busy to write. Sent box express yesterday." On keeping up with the times: Western Union's telegraph stamp instructs, "Be modern—telegraph."[4]

"What had once been considered an impractical invention the government refused to buy for $100,000, had by 1889, in private hands become indispensable to social an economic life."[5] In 1872, the American District Telephone Company was established with four subscribers. It offered telegraphic communication from within the convenience of one's own home for a limited number of services, including the ability to instantly summon a messenger to attend to one's needs. In less than two years, the company opened twelve additional offices and had 2,000 subscribers.

One final illustration of how captivated the American psyche was with telegraphy is to look at it from the viewpoint of literature and language, since they are both reflective of a society's norms and values. In 1882, W.J. Johnson compiled a collection of such literature which deals with a variety of themes from love to burglary to intrigue. His examples: "Kate, an Electro-mechanical Romance," "Song of the Wire," and "The Telegraphers Song."[6]

New words were also introduced into the English language—"telegram" came to replace telegraphic dispatch and "teletype" was a printed rather than written message—and most of all, the "Morse code." *Telegraph and Telephone Age* was of the opinion that the code was truly a new language: "Communications between people was centuries old before Morse was born, but for all those centuries no real progress in the art had been made, either in means or methods. Some acceleration there had been, of course, but none of actual consequence. Then Morse at one stroke, broke through space and pointed the way to what the world enjoys today and to the greater facilities it will have tomorrow."[7]

Chapter Notes

Preface

1. James D. Reid, *The Telegraph in America: Its Founders, Promoters, and Noted Men*, p. 78.
2. Ibid., p. 82.

Introduction

1. James D. Reid, *The Telegraph in America*, p. 110.
2. *Parliament: House of Commons (1899), Documents Relating to the Proposed Pacific Cable*, Ottowa, Canada, p. 9.
3. The main factors were the cost of poles, wire, and wages.
4. Robert Luther Thompson, *Wiring a Continent: The History of the Telegraph Industry in the United States, 1832–1866*, p. 275. In 1869, Western Union President William Orton described the earlier problems of "unity and dispatch in conducting the telegraph business ... the public failed to secure everywhere the benefits of direct and reliable communication. Telegraphic correspondence between the Eastern, Western, and Southern sections was not only burdened with several tariffs but with unnecessary delays [for] copying and retransmission at the termini of each local line.... Another serious evil which the system had to contend with was the existence of competing lines upon the more important routes. The effect ... is to augment the expenses without increasing the business." Orton's conclusion: "Practical men saw that there was but one remedy for these difficulties, and that was by a consolidation of all the rival interests into one organization." Annual Report of the President of Western Union Telegraph Company to the Stockholders, July 13, 1869.

Chapter 1

1. New York *Tribune*, August 9, 1842.
2. An association founded in 1829 for the encouragement of agriculture, commerce, manufacturing, and the arts.
3. New York *Herald*, October 12, 1842.
4. New York *Herald*, October 19, 1842.
5. Ibid.
6. *Samuel F.B. Morse: His Letters and Journals*, Edward Lind Morse, ed. (New York: Houghton Mifflin, 1914), p. 183.
7. Reprinted in the New York *Observer*, December 24, 1842.
8. *Journal of Commerce*, February 3, 1844.
9. New York *Observer*, May 25, 1844.
10. New York *Herald*, May 27, 1844.
11. *National Reporter*, May 28, 1844.
12. New York *Daily Times*, September 14, 1852.
13. *National Police Gazette*, May 30, 1864.
14. New York *Sun*, November 3, 1847.
15. Utica *Gazette*, July 8, 1853.
16. Philadelphia *North American*, January 15, 1846.
17. *Christian Observer*, March 20, 1846.
18. New York *Herald*, May 1, 1845.
19. New York *Herald*, August 3, 1846.
20. *Southern Standard*, May 23, 1846.
21. *Home Journal*, February 4, 1848.
22. Washington *Union*, May 1, 1845.
23. New York *Evening Mirror*, December 15, 1846.

24. *Journal of Commerce*, January 31, 1848.
25. *Mechanics Magazine*, October 8, 1845.
26. Samuel F.B. Morse Papers, Library of Congress, 33, fr. 365.
27. London *Pictorial Times*, October 11, 1845.
28. Samuel F.B. Morse Papers, Library of Congress, 11, fr. 307-308.
29. New York *Observer*, January 16, 1846.
30. Washington *Union*, June 27, 1846.
31. New York *Herald*, October 9, 1845.
32. New York *Tribune*, October 6, 1848.
33. *Day Book*, February 17, 1849.
34. Louisville *Morning Courier*, February 15, 1848.
35. New York *Daily Post*, September 30, 1847.
36. Louisville *Journal*, December 7, 1847.
37. Frankfort *Yeoman*, September 14, 1848.
38. Louisville *Courier*, October 14, 1848.
39. Rochester *Daily Democrat*, March 24, 1849.
40. New York *Sun*, May 6, 1848.
41. New York *Times*, September 14, 1852.
42. Samuel F.B. Morse Papers, Library of Congress, M33, fr. 559.
43. New York *Tribune*, June 1, 1855.
44. New York *Observer*, June 7, 1855.
45. New York *Observer*, May 24, 1855.
46. Ibid.
47. Edinburgh *News of the Churches*, June 21, 1855.
48. New York *Herald*, November 17, 1855.
49. London *Times*, October 13, 1856.
50. Covered in Chapter 5, "The Atlantic Cable."
51. Samuel F.B. Morse Papers, M33, fr. 727-730.
52. *Western Episcopalian*, September 10, 1858.
53. New York *Times*, September 9, 1858.
54. Kenneth Silverman, *Lightning Man: The Accursed Life of Samuel F.B. Morse*, p. 397.
55. *Evening Post*, February 14, 1863.
56. West Virginia *Intelligence*, November 23, 1863.
57. New York *Times*, March 29, 1863.
58. *The North American Review*, January 1864.
59. New York *Tribune*, December 5, 1865.
60. Ibid.
61. New York *Express*, November 5, 1864.
62. New York *Herald*, November 18, 1870.
63. New York *Times*, December 30, 1863.
64. New York *Herald*, December 30, 1868, and January 1, 1869.
65. Kenneth Silverman, *Lightning Man*, p. 438.
66. New York *Times*, April 3, 1872.
67. Louisville *Courier-Journal*, May 3, 1872.
68. *Patent Rights Gazette*, May 16, 1872.
69. New York *Herald*, April 16, 1872.
70. New York *Golden Age*, April 18, 1872.

Chapter 2

1. David Paull Nickles, *Under the Wire: How the Telegraph Changed Diplomacy* (Cambridge: Harvard University Press, 2003), p. 19.
2. Bradford Perkins, *Prologue to War: England and the United States, 1805-1812* (Berkeley: University of California Press, 1961), pp. 431-432.
3. For example, it took approximately one month to travel between England and the United States in the 1800s before ships became powered by steam.
4. Rush to Buchannan #17, March 4, 1848, M34, RG 59 National Archives and Records Administration (NARA).
5. Davis to Washburne, September 6, 1870, M77, Reel 58, RG 59, NARA.
6. David Paull Nickles, "Telegraph Diplomats: The United States' Relations with France in 1848 and 1870," *Technology and Culture*, Vol. 40, No. 1 (January 1999), p. 10.
7. Wickham Hoffman, *Camp, Court, and Siege: A Narrative of Personal Adventure and Observation during Two Wars: 1861-1865, 1870-1871* (New York: Harper and Brothers, 1877), p. 137.

Notes—Chapter 2

8. Daniel R. Headrick, *The Invisible Weapon: Telecommunications and International Politics, 1851-1945* (New York: Oxford University Press, 1991), p. 68.

9. Valerie Cromwell and Zara S. Steiner, "The Foreign Office Before 1914: A Study in Resistance," in *Studies in the Growth of Nineteenth-century Government*, ed. Gillian Sutherland (Totowa, NJ: Rowman and Littlefield, 1972), p. 184.

10. John Kenneth Galbraith, *A Life in Our Times: Memoirs* (Boston: Houghton Mifflin, 1981), pp. 436-437.

11. Even the British foreign office, with much more money at its disposal, discouraged the use of the telegraph. This was especially true for British diplomats stationed in East Asia, where the cost of telegrams to London was exorbitant.

12. David Paull Nickles, "Telegraph Diplomats: The United States' Relations with France in 1848 and 1870," p. 16.

13. At this time (1870), the State Department encouraged diplomatic agents not to transmit telegrams about events that newspapers were reporting in detail.

14. *New York Times*, July 16, 1870.

15. Rachel West, *The Department of State on the Eve of the First World War* (Athens: University of Georgia Press, 1978), pp. 132-133.

16. Deposition of W.H. Seward, 7 July 1870, RG 123 (Records of the U.S. Court of Claims), General Jurisdiction, Case File 6151, Box 306, NARA.

17. David Kahn, *The Codebreakers: The Story of Secret Writing* (New York: Macmillan, 1967), p. 189.

18. Wickham Hoffman to Hamilton Fish, July 18, 1870.

19. The Hohenzollerns were a German noble family.

20. Georges Bonnin, ed., *Bismarck and the Hohenzollern Candidate for the Spanish Throne: The Documents in the German Diplomatic Archives* (London: Chatto and Windus, 1957), pp. 233-234.

21. Vary T. Coates, et al., *A Retrospective Technology Assessment: Submarine Telegraphy—The Transatlantic Cable of 1866* (San Francisco: San Francisco Press, 1979), p. 87.

22. Ibid., pp. 83-87.

23. Elting Morison, *Men, Machines, and Modern Times* (Cambridge, MA: MIT Press, 1966), pp. 36-38.

24. Yakup Bektas, "The Sultan's Messenger: Cultural Constructions of Ottoman Telegraphy, 1847-1880," *Technology and Culture*, Vol. 41, No. 4 (October 2000), p. 670.

25. A pasha is a high ranking Ottoman official, usually a governor.

26. Cyrus Hamlin, *Among the Turks* (New York: R. Carter and Bros., 1878), pp. 185-195.

27. Viscount Stratford de Redcliffe, British ambassador to the Ottoman empire, to Earl Clarendon, British secretary of state for foreign affairs, September 1, 1857.

28. The official name of the British company formed to build the telegraph link between Britain and India.

29. London *Times*, September 16, 1858.

30. Warren Frederick Ilchman, *Professional Diplomacy in the United States, 1779-1939: A Study in Administrative History* (Chicago: University of Chicago Press, 1961), p. 19.

31. Edward John Phelps, *International Relations: Address Before the Phi Beta Kappa Society of Harvard University, June 29, 1889* (Burlington, VT: Free Press Association, 1889), p. 11.

32. Donald E. Queller, *The Office of Ambassador in the Middle Ages* (Princeton: Princeton University Press, 1967), p. 228.

33. Ralph E. Weber, *United States Diplomatic Codes and Ciphers, 1775-1938* (Chicago: Precedent, 1979), p. 154; and Sir Herbert Maxwell, *The Life and Letters of George William Frederick, Fourth Earl of Clarendon* (London: Edward Arnold, 1913), p. 1718.

34. P.M. Kennedy, "Imperial Cable Communications and Strategy, 1870-1914," *English Historical Review*, 86 (1971), p. 751.

35. Monteagle Stearns, *Talking to Strangers: Improving American Diplomacy at Home and Abroad* (Princeton: Princeton University Press, 1996), p. 116.

36. John A. Munroe, *Louis McLane: Federalist and Jacksonian* (New Brunswick:

Rutgers University Press, 1973), pp. 533–534.

37. "The Need of a Trained Diplomatic Service," *New York Times*, April 29, 1900, p. 25.

38. "Diplomacy as a Profession," *National Review* (London) 35 (March 1900), p. 101.

39. Gerald G. Eggert, *Richard Olney: Evolution of a Statesman* (University Park: Pennsylvania State University Press, 1974), p. 318.

40. J. D. Gregory, *On the Edge of Diplomacy: Rambles and Reflections, 1902–1928* (London: Hutchinson, 1931), p. 15.

41. David Paull Nickles, *Under the Wire: How the Telegraph Changed Diplomacy*, p. 46.

42. Finley Peter Dunne, "Mr. Dooley on Diplomatic Indiscretions," New York Times, March 22, 1914, sec. 5, 5.

43. Viscount Stratford de Redcliffe, May 13, 1861, in *British Parliamentary Papers, Report from the Select Committee on Diplomatic Service, Reports from Committees* 6 (1861), p. 168.

44. John Tilly and Stephen Gaselee, *The Foreign Office* (London: Putnam's, 1933), p. 257.

45. Paleologue was correct. This war plan of Germany's was known as the Schlieffen Plan, conceived by Count Alfred von Schlieffen, who realized that since Germany faced a two-front war—Russia to the East and France to the West, Germany needed to attack and defeat France as quickly as possible and then turn and attack Russia.

46. J.F.V. Keiger, *Raymond Poincare* (Cambridge: Cambridge University Press, 1997), p. 176.

47. Baron Wilhelm von Schoen, *The Memoirs of an Ambassador: A Contribution to the Political History of Modern Times*, trans. Constance Vesey (London: George Allen and Unwin, 1922), pp. 200–201.

48. David Paull Nickles, *Under the Wire: How the Telegraph Changed Diplomacy*, pp. 50–52.

49. C.W. de Kiewiet, *The Imperial Factor in South Africa: A Study in Politics and Economics* (Cambridge: Cambridge University Press, 1937), p. 293.

50. Lugi Albertini, *The Origins of the War of 1914*, Vol. 2 (Oxford: Oxford University Press, 1953), p. 504.

51. Andy Smith, "Sun Sets on 150 Years of the Foreign Office Cable," *Observer*, August 16, 1998, p. 6.

52. Banquet speech, April 15, 1864, in *Europe and America: Report on the Proceeding at an Inauguration Banquet Given by Mr. Cyrus Field* (London: William Brown, 1864), Box 142 (Cyrus Field papers), Collection #205 (Western Union Telegraph Company), 1993 Addendum, National Museum of American History.

53. Raymond A. Jones, *The British Diplomatic Service, 1815–1914* (Gerrards Cross, U.K.: Colin Smythe, 1983), p. 125.

54. "Intellectual Effects of Electricity," *Spectator*, Nov. 9, 1889.

55. "Cipher and Cipher Keys," *Living Age* 333 (Sept. 15, 1927), p. 493.

56. Svante Lindqvist, "Changes in the Technological Landscape: The Temporal Dimension in the Growth and Decline of Large Technological Systems," in *Economics of Technology*, ed., O. Grandstand (Stockholm, 1994), pp. 271–288.

57. David Paull Nickles, "Telegraph Diplomats: The United States' Relations with France in 1848 and 1870," *Technology and Culture*, Vol. 40, No. 1 (January 1999), p. 24.

Chapter 3

1. David Homer Bates, *Lincoln in the Telegraph Office: Recollections of the United States Military Telegraph Corps* (Lincoln: University of Nebraska, 1995), pp. 31–32.

2. Thomas C. Jepsen, "Women Telegraph Operators on the Western Front," *Journal of the West* 35 (2): 1996, p. 75.

3. Francis Miller, *The Photographic History of the Civil War* (New York: Review of Reviews Co., 1911), p. 344.

4. William R. Plum, *The Military Telegraph During the Civil War in the United States* (New York: Arno Press, 1974), p. 96.

5. This was the semaphore system.

Notes—Chapter 3

6. Paul J. Scheips, "Union Signal Communications: Innovation and Conflict," *Civil War History*, Vol. 9, No. 4 (December 1963), p. 22.

7. This opinion changed when McClellan's chief quartermaster and chief commissary officer saw its usefulness in the field.

8. Robert Thompson, *Wiring a Continent* (Princeton: Princeton University Press, 1947), p. 385.

9. Stephen W. Sears, *Chancellorsville* (Boston: Houghton Mifflin, 1996), p. 196.

10. William Plum, *The Military Telegraph During the Civil War in the United States*, pp. 125-126.

11. *The War of the Rebellion: A Compilation of the Official Records of the Union and Confederate Armies*, Series 1, 17, Part 2, pp. 385-386.

12. James H. Wilson, *Under the Old Flag* (New York: D. Appleton, 1912), 1, p. 396.

13. *The War of the Rebellion: A Compilation of the Official Records of the Union and Confederate Armies*, 45, Part 1, p. 1172; 17, Part 2, pp. 385-386, 394, 379.

14. Jesse Bowman Young, *The Battle of Gettysburg* (New York: Harper, 1913), p. 296.

15. *The War of the Rebellion. A Compilation of the Official Records of the Union and Confederate Armies*, 36, Part 3, p. 321. Major Eckert, assistant superintendent of military telegraphers, in his report actually praised Caldwell. Ibid., 51, Part 1, p. 200.

16. Lew Wallace, *An Autobiography* (New York: Harper and Brothers, 1906), 2, p. 783.

17. Jacob D. Cox, *The Battle of Franklin* (New York: Charles Scribner's Sons, 1897), pp. 168-169.

18. David Homer Bates, *Lincoln in the Telegraph Office*, pp. 313-315.

19. John M. Schofield, *Forty-six Years in the Army* (Norman: University of Oklahoma, 1998), p. 206.

20. LaFayette C. Baker, *History of the United States Secret Service* (Philadelphia: L.C. Baker, 1867), pp. 124-126.

21. When a telegram is sent, the time it was transmitted and the time received are stamped on it.

22. *The War of the Rebellion: A Compilation of the Official Records of the Union and Confederate Armies*, 17, Part 2, pp. 378-379.

23. Ibid., p. 347.

24. Ibid., p. 370.

25. John M. Schofield, *Forty-six Years in the Army*, pp. 232-233.

26. Jacob Cox, *Military Reminiscences of the Civil War* (New York: Charles Scribner's Sons, 1900), pp. 135-137.

27. Horace Porter, *Campaigning with Grant* (Alexander, VA: Time-Life Books, 1981), pp. 358-359.

28. John Keegan, *Mask of Command* (New York: Elisabeth Sifton Books, 1987), p. 210.

29. Charles D. Ross, *Trial by Fire* (Shippensburg, PA: White Mane Books, 2000), p. 160.

30. This was the same Caldwell who, because he was scared, left General Meade's command at Gettysburg and moved twenty-five miles away to a safe position with the military codes.

31. William R. Plum, *The Military Telegraph During the Civil War in the United States*, pp. 352-355. These letters from Union generals during the war were sent to William Plum in response to his letters to them asking for their opinion of the military telegraph.

32. John Schofield, *Forty-six Years in the Army*, pp. 474-476.

33. Roscoe Pound, "The Military Telegraph in the Civil War," *Proceedings of the Massachusetts Historical Society*, Third Series, Vol. 66 (October 1936-May 1941), pp. 200-203.

34. A telegraph sounder is the part of the instrument that produces the "click" so that messages can be heard.

35. D.H. Hill to Morris, May 8, 1863, and Jno. R. Trucks to Morris, October 11, 1863, both in the Southern Telegraph Company Papers.

36. R.H. Glass to Morris, May 4, 1863, in Southern Telegraph Papers.

37. J. Cutler Andrews, "The Southern Telegraph Company, 1861-1865: A Chapter in the History of Wartime Communica-

tion," *The Journal of Southern History*, Vol. 30, No. 3 (August 1964), p. 335.

38. Charleston *Mercury*, December 2, 1861.

39. Morris to Brigadier General Isaac Munroe St. John, March 22, 1864, in Southern Telegraph Papers. In this letter Morris told St. John (who was the Confederate commissary general): "It has become necessary to make Blue Stone for the Batteries in use on the Telegraph Lines. The supply in the Confederate States is almost exhausted."

40. Savannah *Republican*, June 29, 1863.

41. A.B. Magruder to Morris, February 24, 1862.

42. J.B. Tree to Morris, November 8, 1864.

43. Morris to Cooper, January 1, 1863.

44. Morris to Seddon, March 3, 1864.

45. Morris to W.P. Miles, chairman, Military Committee, House of Representatives, January 20, 1864, in National Archives Microfilm Publications, Confederate Papers Relating to Citizens or Business Firms.

46. R.E. Lee to Morris, November 14, 1862.

47. Mobile *Register and Advertiser*, October 8, 1864. In the *Journal of the Telegraph*, May 1, 1868, p. 4, is a news item about the indictment of one Wm. Roche, a telegraph operator in New York City, for divulging a telegraph message to John Sammond, a Wall Street broker.

48. J.B. Magruder to S. Cooper, November 17, 1861.

49. William Plum, *The Military Telegraph During the Civil War in the United States*, 2, pp. 118-119.

50. Francis B.C. Bradlee, *Blockade Running During the Civil War, and the Effect of Land and Water Transportation on the Confederacy* (Salem, MA: Essex Institute Press, 1925), p. 307.

51. J.M. Crowley to Morris, November 2, 1863.

52. Richmond *Daily Dispatch*, October 3, 1862.

53. William R. Plum, *The Military Telegraph During the Civil War in the United States*, 2, pp. 247-248. Morris wrote to John S. Preston, chief of the Bureau of Conscription, on January 11, 1864, that it was now impossible to get telegraph wire from abroad.

54. Ibid., p. 249.

55. Tree to Morris, December 6, 1864.

56. Tree to Morris, March 7, 1865.

57. John Emmett O'Brien, *Telegraphing in Battle* (Wilkes-Barre, PA: Reader Press, 1910), pp. 295-296.

Chapter 4

1. William R. Plum, *The Military Telegraph During the Civil War in the United States* (New York: Arno Press, 1974), p. 77.

2. David Homer Bates, *Lincoln in the Telegraph Office* (Lincoln: University of Nebraska Press, 1995), p. 20.

3. Tom Wheeler, *Mr. Lincoln's T-mails* (New York: Harper Collins, 2006), p. 7.

4. David Homer Bates, *Lincoln in the Telegraph Office*, p. 7.

5. Ibid., p. 282.

6. Roy Basler, ed., *The Collected Works of Abraham Lincoln* (New Brunswick: Rutgers University Press, 1953), p. 87.

7. Ibid., p. 90.

8. Ibid., p. 91.

9. Ibid., pp. 91-92.

10. E.B. Long, *The Civil War Day by Day* (Garden City, NY: Doubleday, 1971), p. 160.

11. Roy Basler, ed. *The Collected Works of Abraham Lincoln*, Vol. 5, p. 203.

12. Benjamin P. Thomas, *Abraham Lincoln: A Biography* (New York: Barnes and Noble, 1994), p. 320.

13. Roy Basler, ed., *The Collected Works of Abraham Lincoln*, Vol. 5, p. 230.

14. Ibid., p. 232.

15. *The War of the Rebellion: A Compilation of the Official Records of the Union and Confederate Armies*, 128 vols., Government Printing Office, 1880-1901, Series 1, Vol. 12 (Part 3), p. 220.

16. Roy Basler, ed. *The Collected Works of Abraham Lincoln*, p. 233.

17. Ibid., p. 232.

18. Ibid., p. 235.

19. Ibid., p. 279.

Notes—Chapter 5

20. Ibid., p. 289.
21. Tom Wheeler, *Mr. Lincoln's T-mails*, p. 65.
22. Stephen W. Sears, *Controversies and Commanders: Dispatches from the Army of the Potomac* (Boston: Houghton Mifflin, 1999), p. 70.
23. Roy Basler, ed., *The Collected Works of Abraham Lincoln*, Vol. 6, p. 22.
24. *The War of the Rebellion: A Compilation of the Official Records of the Union and Confederate Armies*, p. 43.
25. Roy Basler, ed., *The Collected Works of Abraham Lincoln*, p. 314.
26. *The War of the Rebellion: A Compilation of the Official Records of the Union and Confederate Armies*, p. 92.
27. David Homer Bates, *Lincoln in the Telegraph Office*, p. 159. The last sentence of the telegram referred to an effort made by the Democratic Party to nominate prominent military men for office throughout the country.
28. Ibid., p. 182.
29. Roy Basler, ed., *The Collected Works of Abraham Lincoln*, p. 53.
30. Library of Congress, the Abraham Lincoln Papers, Grant Papers.
31. *The War of the Rebellion: A Compilation of the Official Records of the Union and Confederate Armies*, pp. 40–42.
32. David Homer Bates, *Lincoln in the Telegraph Office*, pp. 244–245.
33. Ibid., p. 122.
34. Ibid., p. 255.
35. Roy Basler, ed., *The Collected Works of Abraham Lincoln*, p. 476.
36. Ibid., pp. 73–74.
37. Wiley Sword, *The Confederacy's Last Hurrah* (Lawrence: University Press of Kansas, 1993), p. 278.
38. Ibid., p. 45.
39. Ibid., p. 11.
40. Lincoln's Report to the House of Representatives, February 18, 1865, on the Hampton Roads Peace Conference.
41. Library of Congress, the Abraham Lincoln Papers.
42. William Plum, *The Military Telegraph During the Civil War in the United States*, pp. 34–35. The author states that by reading Lincoln's cipher backward with regard to the phonetics rather than the orthography, the meaning will be apparent.
43. Roy Basler, ed., *The Collected Works of Abraham Lincoln*, p. 392.
44. John Emmet O'Brien, *Telegraphing in Battle* (Scranton: Raeder Press, 1910), p. 260.
45. David Homer Bates, *Lincoln in the Telegraph Office*, p. xxi.
46. Roy Basler, ed., *The Collected Works of Abraham Lincoln*, p. 43.
47. Ibid., p. 231.
48. Roy Basler, ed., *The Collected Works of Abraham Lincoln*, p. 347.

Chapter 5

1. Samuel J. Prime, *The Life of Samuel F.B. Morse* (New York: Arno Press, 1974), p. 616.
2. The same plan that Daniel Craig had.
3. Maury to Dobbin, February 22, 1854.
4. Peter Cooper, "Reminiscences."
5. Cyrus Field, *History of the Atlantic Telegraph* (Freeport, NY: Books for Libraries, 1972), pp. 41–42.
6. Samuel Carter III, *Cyrus Field: Man of Two Worlds* (New York: G.P. Putnam's Sons, 1968), p. 110.
7. Cyrus Filed, *The Story of the Atlantic Telegraph* (New York: Scribner's, 1892), p. 80.
8. Chester G. Hearn, *Circuits in the Sea: The Men, the Ships, and the Atlantic Cable* (Westport, CT: Praeger, 2004), p. 48.
9. John Merrett, *Three Miles Deep: The Story of the Transatlantic Cables* (London: Hamish Hamilton, 1958), p. 52.
10. Isabella Field Judson, *Cyrus W. Field, His Life and Work* (New York: Harper and Brothers, 1896), p. 77.
11. Kenneth Silverman, *Lightning Man: The Accursed Life of Samuel F.B. Morse* (New York: Alfred A. Knopf, 2003), p. 354.
12. Ibid., p. 355.
13. Bern Dibner, *The Atlantic Cable* (Norwalk, CT: Burndy Library, 1959), p. 24.
14. New York *Tribune*, August 27, 1857.
15. *Scientific American*, October 3, 1857.

16. Samuel Morse to Sidney Morse, March 15, 1858.

17. John Mullaly, *The Laying of the Cable or the Ocean Telegraph, Being a Complete and Authentic Narrative of the Attempt to Lay the Cable* (New York: D. Appleton, 1858), p. 284.

18. London *Times*, August 11, 1858.

19. Charles Bright, *The Story of the Atlantic Cable* (New York: D. Appleton, 1903), p. 149.

20. London *Times*, September 6, 1858.

21. Cyrus Field, *The History of the Atlantic Telegraph*, p. 246.

22. Kenneth Silverman: *The Accursed Life of Samuel F.B. Morse*, pp. 377–378.

23. Cyrus' brother.

24. Cyrus Field, *History of the Atlantic Telegraph*, p. 252.

25. Cyrus Field, *History of the Atlantic Cable*, p. 269.

26. Reuter to Field, November 19, 1862, Field Papers.

27. Isabella Field Judson, *Cyrus Field*, p. 172.

28. Anderson to Field, November 3, 1865.

29. Field to Associated Press, August 28, 1866.

30. John Merrett, *Three Miles Deep*, p. 150.

31. Chester G. Hearn, *Circuits in the Sea*, p. 239.

32. Ibid. p. 248.

Chapter 6

1. James D. Reid, *The Telegraph in America* (New York: Derby, 1879), p. 359.

2. "Sketch of the Life of J.H. Wade from 1811 to About 1867," Jepha H. Wade Papers, Series 1, Container 1, Folder 2, p. 15, Western Reserve Historical Society, Cleveland, Ohio.

3. W.J. Johnson, *Telegraph Tales and Telegraph History* (New York: W. J. Johnson, 1880), pp. 181–182.

4. *Fifth Annual Report of the Pennsylvania Railroad Company to the Stockholders*, pp. 63–64.

5. "Report of the Directors of the New York and Erie Railroad to the Stockholders," *American Railroad Journal*, Vol. 27, January 14, 1854, p. 21.

6. Steven W. Usselman, *Regulating Railroad Innovation: Business, Technology, and Politics in America* (Cambridge: Cambridge University Press, 2002), p. 126.

7. Annteresa Lubrano, *The Telegraph, How Technology Innovation Caused Social Change*, pp. 44–45.

8. Charles Francis Adams, *Notes on Railroad Accidents* (New York: G. P. Putnam's Sons, 1879), p. 140.

9. Ibid., pp. 64–65. The Revere accident resulted in eighty-six casualties, including twenty-nine deaths.

10. Benjamin Sidney Michael Schwantes, "Fallible Guardian: The Social Construction of Railroad Telegraphy in 19th Century America." Dissertation, University of Delaware, Fall 2008, p. 176.

11. Ibid., p. 177.

12. An honor.

13. Kenneth Silverman, *Lightning Man: The Accursed Life of Samuel F.B. Morse*, pp. 182–183.

14. Benjamin Sidney Michael Schwantes, "Fallible Guardian," p. 185.

15. Alfred D. Chandler, Jr., *The Visible Hand: The Managerial Revolution in American Business* (Cambridge: Harvard University Press, 1977), p. 103.

16. Thomas P. Hughes, *American Genesis* (New York: Viking, 1989), p. 4.

17. Ibid., p. 3.

18. Alfred D. Candler, *The Visible Hand*, p. 195.

19. James D. Reid, *The Telegraph in America*, p. 480.

20. Ibid.

21. Ibid., pp. 480–481.

22. Robert L. Thompson, *Wiring a Continent*, p. 216.

23. Annteresa Lubrano, *The Telegraph: How Technology Innovation Caused Social Change*, pp. 48–49.

Chapter 7

1. "The Electric Telegraph," *Daily Chronicle*, November 16, 1847.

Notes—Chapter 7

2. "Influence of the Telegraph on Commerce," *Hunt's Merchants' Magazine*, 59 (August 1868), pp. 106-107.

3. "The Atlantic Telegraph," *The Commercial and Financial Chronicle*, 1 (July 8, 1865), 34.

4. "The Magnetic Telegraph West," *Public Ledger and Daily Transcript* (Philadelphia), August 24, 1846.

5. "The Telegraph," *Republican* (St. Louis), November 23, 1847.

6. *Report of the Proceedings of a Meeting of the Stockholders of the Atlantic and Ohio Telegraph Company*, July 19, 1849.

7. Alfred D. Chandler, Jr., *The Visible Hand*, p. 218.

8. Fred M. Jones, "The Middleman in the Domestic Trade of the United States, 1800-1860," *Illinois Studies in Social Sciences*, 21, No. 3 (Urbana, IL, 1937), p. 15.

9. S.S. Huebner, "The Functions of Produce Exchanges," *Annals of the American Academy of Political and Social Science*, 38 (September 1911), pp. 319-353.

10. Morton Rothstein, "Antebellum Wheat and Cotton Exports: A Contrast in Marketing Organization and Economic Development," *Agricultural History*, 40 (April 1966), p. 96.

11. United States Bureau of Corporations, *Report of the Commissioner of Corporations on the Beef Industry*, March 3, 1905 (Washington, D.C., 1905), p. 207.

12. Harold Woodman, *King Cotton and His Retainers: Financing and Marketing the Cotton Crop of the South, 1800-1925* (Lexington: University of Kentucky Press, 1968), pp. 288-289.

13. Alfred D. Chandler, *The Visible Hand*, p. 316.

14. R.B. DuBoff, "Business Demand and the Development of the Telegraph in the United States, 1844-1860," *Business History Review*, 54 (Winter 1980), pp. 461-462.

15. Glenn Porter, *The Rise of Big Business, 1860-1910* (New York: John Wiley and Sons, 1973), p. 43.

16. Alfred Vail, *The American Electro Magnetic Telegraph with the Reports of Congress, and a Description of All Telegraphs Known* (Philadelphia: Lea and Blanchard, 1845), p. 49.

17. Du Pont de Nemours Powder Company, *Private Telegraph Code 1905* (Wilmington, DE: Dando, 1905), pp. 58, 194.

18. Thomas Cochran and William Miller, *The Age of Enterprise: A Social History of Industrial America*, rev. ed. (New York: Harper, 1961), p. 66.

19. James D. Reid, *The Telegraph in America*, p. 596.

20. George Prescott, *Electricity and the Electric Telegraph* (New York: D. Appleton, 1877), p. 682.

21. Ibid., pp. 621-626.

22. United States Congress, Senate, *Investigation of Western Union and Postal Telegraph Companies*, 60th Congress, 2d session, S. Doc. 725 (Washington, D.C., 1909), pp. 21-22.

23. "The Southern Telegraph," *New Orleans Commercial Times*, July 13, 1847.

24. J.S. Gibbons, *The Banks of New-York, Their Dealers, the Clearing House, and the Panic of 1857* (New York: D. Appleton, 1858), pp. 356-357.

25. D.A. Wells, *Recent Economic Changes and Their Effect on the Production and Distribution of Wealth and the Well-Being of Society* (New York: D. Appleton, 1899), p. 82.

26. Alvin Harlow, *Old Wires and New Waves: The History of the Telegraph, Telephone, and Wireless* (New York: D. Appleton, 1936), pp. 150-151.

27. James D. Reid, *The Telegraph in America*, pp. 168-169. In addition to his position with the Magnetic Telegraph Company, Reid was also a superintendent with both the Lake Erie and the Pittsburgh, Cincinnati, and Louisville telegraph companies. His book, written in 1879, was the most detailed history of the telegraph published at that time.

28. Richard B. du Boff, "The Telegraph in Nineteenth-century America: Technology and Monopoly," *Comparative Studies in Society and History*, Vol. 26, No. 4 (October 1984), p. 581.

Notes—Chapter 8

Chapter 8

1. Robert L. Thompson, *Wiring a Continent*, p. 235.
2. Even though the establishment of the Postal Telegraph Company satisfied the opponents of monopolies at that time, the Postal Telegraph Company was decidedly junior to Western Union in all respects. The telegraph business was henceforth synonymous with Western Union.
3. William Orton, *Government Telegraphs: Argument of William Orton, President of the Western Union Telegraph Company on the Bill to Establish Postal Telegraph Lines, Delivered Before the Select Committee of the United States House of Representatives* (New York: Russells American Steam Printing House, 1870.
4. *New York Herald*, May 23, 1846.
5. Martin Van Creveld, *Technology and War* (New York: Free Press, 1989), p. 53.
6. James D. Reid, "The Western Press and the Telegraph," *National Telegraph Review* 1 (October 1853), pp. 234–238.
7. *Rochester Democrat*, O'Rielly Telegraph Collection, 1st Series, Vol. 2.
8. William F.S. Shanks, "How We Get Our News," *Harpers New Monthly Magazine* (May 1967), p. 514.
9. Ibid., p. 514.
10. *New York World*, January 20, 1884.
11. *Louisville Journal*, November 6, 1847.
12. James D. Reid, "The Western Press and the Telegraph," p. 233.
13. *Philadelphia North American*, December 1845.
14. Menahem Blondheim, *News Over the Wires: The Telegraph and the Flow of Public Information in America, 1844–1897* (Cambridge: Harvard University Press, 1994), p. 67.
15. The description of this incident is based on records of *Daniel H. Craig v. Francis O.J. Smith* in the Superior Court of New York, and of *Commonwealth of Massachusetts v. Francis O.J. Smith*, first in the Superior Court, Suffolk County, then in the Supreme Judicial Court of Massachusetts.
16. *New York Herald*, August 19, 1846.
17. Menahem Blondheim, *News Over the Wires: The Telegraph and the Flow of Public Information in America, 1844-1897*, p. 5.
18. *New York Sun*, November 25, 1846.
19. Menahem Blondheim, *News Over the Wires: The Telegraph and the Flow of Public Information in America, 1844–1897*, pp. 71–78.
20. *Pittsburgh Mercantile Advertiser*, July 8, 1848.
21. "An Act to Provide for the Incorporation and Regulation of Telegraph Companies," April 12, 1848, New York *Laws*, 1848.
22. Daniel Craig, printed circular to managers of telegraph companies (May 1851).
23. Smith to O'Reilly, September 2, 1852.
24. New York *Courier and Enquirer*, July 13, 1852.
25. Daniel Craig, *Reply of the Executive Committee to the Pamphlet of D.H. Craig, Agent of the New York Associated Press* (New York, 1860), p. 3.
26. Oliver Gramling, *AP: The Story of News* (New York: J. J. Little and Ives, 1940), p. 31.
27. Ibid., p. 39.
28. Ibid., p. 61.
29. Edwin Emery, *The Press and America: An Interpretative History of Journalism*, 2nd ed. (Englewood Cliffs, NJ: Prentice-Hall, 1962), p. 317.
30. U.S. Congress, House of Representatives. 41st Congress, 2nd Session (1869–1870). House Report No. 114 (serial volume 1438), *The Postal Telegraph in the United States* (Washington, D.C.: Government Printing Office, 1870), p. 51.
31. Fred S. Siebert, Theodore Paterson, and Wilbur Schramm, *Four Theories of the Press* (Urbana: University of Illinois Press, 1956), p. 60.
32. Oliver Grambling, *AP: The Story of News* (New York: J.J. Little and Ives, 1946), p. 67.
33. *The Postal Telegraph in the U.S.*, p. 47.
34. *Postal Telegraph System*, p. 15.
35. *Annual Report of the President of the Western Union Telegraph Company*, 1880 (pp. 9–10), and 1887 (pp. 10–11).

36. U.S. Bureau of the Census, *Statistical Abstract of the United States, 1901* (Washington: Government Printing Office, 1902), p. 404.

37. U.S. Congress, House of Representatives, 42nd Congress, 3rd Session (1872–73). House Miscellaneous Document No. 73 (serial number 1572), *Postal Telegraph* (Washington: Government Printing Office, 1873), p. 26.

38. *Postal Telegraph System*, p. 2.

39. Peter R. Knights, "Financial History of the Boston Evening *Transcript* 1866–1900," unpublished study, University of Wisconsin, 1965.

40. At the same time, the price of newsprint also declined, freeing up more money with which to purchase wire news coverage.

41. Victor Rosewater, *History of Cooperative News-gathering in the United States* (New York: D. Appleton, 1930), passim.

42. *Annual Report of the President of the Western Union Telegraph Company* (1888), p. 10.

43. Alfred McClung Lee, *The Daily Newspaper in America: The Evolution of a Social Instrument* (New York: Macmillan, 1937), p. 526.

44. Wisconsin Press Association, *Proceedings*, 1872, p. 40.

45. Ibid., pp. 21–22.

46. Ibid., p. 16.

47. Donald L. Shaw, "News Bias and the Telegraph: a Study of Historical Change," *Journalism Quarterly*, Spring 1967, p. 10.

Chapter 9

1. Samuel Irenaeus Prime, *The Life of Samuel F.B. Morse, LL.D* (New York: D. Appleton, 1875), p. 508.

2. Frank G. Carpenter, "Henry Clay on Nationalizing the Telegraph," *The North American Review*, Vol. 154, No. 424 (March 1892), pp. 381–382.

3. Annteresa Lubrano, *The Telegraph: How Technological Innovation Caused Social Change* (New York: Garland, 1997), p. 70.

4. Federal Communications Commission, "Corporate History of the Western Union Telegraph Company," 1915; 73d Congress, 2d Session, House Report No. 1273, Part 3, No. 4 (1935) 3967–74. Referred to as the Splawn Report.

5. Horace Coon, *American Tel. & Tel.: The Story of a Great Monopoly* (New York: Longmans, Green, 1939), p. 31.

6. Congressional Globe, 39th Congress, 1st Session (1866) pp. 979–980.

7. See testimony of President William Orton of Western Union, 41st Congress, 2d Session, House Report No. 114 (July 5, 1870), p. 118.

8. Annual Report of the President of the Western Union Telegraph Company, 1869, p. 9.

9. Edison to Wiman, July 5, 1884.

10. Annual Report of the President of Western Union, 1893, p. 7.

11. Ibid.

12. Ibid., 1900, p. 7.

13. Frank Parson, *The Telegraph Monopoly* (Philadelphia: C.F. Taylor, 1899), pp. 11–13.

14. Splawn Report, 3979; Annual Report of the President of Western Union, 1889–1899; H.H. Goldin, Governmental Policy and the Domestic Telegraph Industry, *The Journal of Economic History*, Vol. 7, No.1 (May 1947), pp. 57–58. At this time, H.H. Goldin was assistant chief, Economics and Statistics Division, Federal Communication Division.

15. David Ames Wells, *The Relation of the Government to the Telegraph* (New York, 1873), p. 12.

16. Ibid.

17. Between 1866 and 1873 eight additional companies were created and became competitors of Western Union.

18. David Ames Wells, *The Relation of the Telegraph to the Government*, p. 13.

19. Ibid., pp. 13–14.

20. Ibid., pp. 20.

21. Herbert N. Casson, *The History of the Telephone* (Chicago: A.C. McClurg, 1910), pp. 58–59.

22. 76th Congress, 1st Session, House Document No. 340 (June 14, 1939) 3, pp. 123–125.

23. Annteresa Lubrano, *The Telegraph:*

Notes—Chapter 10

How Technological Innovation Caused Social Change, p. 17.

24. David Hochfelder, *The Telegraph in America, 1832-1890* (Baltimore: Johns Hopkins University Press, 2012), p. 222.

25. Michael J. Makley, *John Mackay: Silver King in the Gilded Age* (Reno: University of Nevada Press, 2009), pp. 175-189.

26. This invention allowed four separate signals to be sent on a single wire. Edison sold the rights to Western Union in 1874 for $10,000.

27. David Hochfelder, *The Telegraph in America, 1832-1890*, pp. 41-42.

28. Ibid., pp. 45-46.

29. A.N. Holcombe, "Public Ownership of Telegraphs and Telephones," *The Quarterly Journal of Economics*, Vol. 28, No. 3 (May 1914), p. 582.

30. Senate Committee on Post Offices and Post Roads, *Report on Postal Telegraph*, 48th Congress, 1st session, 1884, S. Report 577, 1 and 5.

31. Gardiner G. Hubbard, "The Proposed Changes in the Telegraphic System," *North American Review*, 117 (July 1873), p. 101.

32. However, by 1869, the Western Union Telegraph Company was generating a profit of $2,748,801.45 (Annual Report to the Stockholders, 1905, p. 6), while the postal service was operating at deficit of over six million dollars, according to an article in the New York *Times* on January 21, 1869. Clearly, the profitability of the company combined with the Post Office Department losses would put the postal service in a better economic position.

33. William Stanley Jevons, *Methods of Social Reform and Other Papers* (London: Macmillan, 1883), p. 293.

34. David Hochfelder, *The Telegraph in America*, p. 53.

35. Jeffrey Kieve, *The Electric Telegraph: A Social and Economic History* (Newton Abbot: David and Charles, 1973), Chapter 9.

36. New York *Times*, January 21, 1869.

37. William E. Nelson, *The Roots of American Bureaucracy* (Cambridge: Harvard University Press, 1982), p. 25.

38. This testimony was taken from the House Committee on Appropriations, April 1871, House of Representatives Bill 1083.

39. Annteresa Lubrano, *The Telegraph: How Technology Innovation Caused Social Change*, pp. 104-107.

40. Edwin Gabler, *The American Telegrapher* (New Brunswick: Rutgers University Press, 1988), p. 46.

41. Ibid., pp. 46-47.

42. Alexander James Field, "The Magnetic Telegraph, Price and Quantity Data, and the New Management of Control," *Journal of Economic History*, Vol. 52, No. 2 (June 1992), p. 413.

43. Annteresa Lubrano, *The Telegraph: How Technology Innovation Caused Social Change*, p. 83.

44. David F. Noble, *America by Design* (New York: Oxford University Press, 1977), p. 15.

45. James D. Reid, *The Telegraph in America* (New York: James D. Reid, 1879), p. 554.

46. Western Union Annual Report to the Stockholders, 1873, p. 19.

47. Tom Standage, *The Victorian Internet* (New York: Walker, 1998), pp. 133-134.

48. W. J. Johnson, *Lightning Flashes and Electric Dashes* (Cambridge: Harvard University Press, 1882), p. 50.

49. Edwin Gabler, *The American Telegrapher*, p. 146.

50. Alice Kessler-Harris, *Out to Work: A History of America's Wage Earning Women* (New York: Oxford University Press, 1982), p. 16.

51. Edwin Gabler, *The American Telegrapher*, p. 57.

52. Lester Lindley, *The Constitution Faces Technology* (New York: Arno Press, 1975), p. 128.

Chapter 10

1. Alvin Harlow, *Old Wires and New Waves*, p. 243.

2. Final Report and Testimony, Commission on Industrial Relations, April 1914, Vol. 10, p. 9428.

3. *The Telegrapher*, November 4, 1864.
4. *Commercial Telegraphers' Journal*, Vol. 13–14, p. 503.
5. Statement of O.H. Palmer, Alonzo Cornell, and Geo. Walker, vice presidents of Western Union, New York *Times*, January 5, 1870, p. 1.
6. New York *Herald*, January 19, 1870.
7. Vidkunn Ulriksson, *The Telegraphers, Their Craft and Their Unions* (Washington, D.C.: Public Affairs Press, 1953), pp. 28–29.
8. Constitution, Brotherhood of Telegraphers, Sec. 1, Article 15, *Commercial Telegraphers' Journal*, May 1914.
9. New York *Herald*, July 17, 1883.
10. New York *Daily Tribune*, July 18, 1883.
11. A spokesman for Western Union.
12. Vidkunn Ulriksson, *The Telegraphers: Their Craft and Their Unions*, p. 34.
13. New York *Herald*, July 20, 1883.
14. *Scientific American*, July 28, 1883.
15. New York *Daily Tribune*, August 19, 1883.
16. Postmaster general's report, 1890.
17. *Electrical World*, October 10, 1891.
18. Ibid., July 14, 1888.
19. Report of Hearings, U.S. Industrial Commission, 1914, p. 144.
20. Frank Parsons was a social reformer, engineer, lawyer, writer, and social commentator. He taught at Kansas State Agricultural College. One of his objectives was the regulation of monopolies.
21. New York *Daily Tribune*, May 10, 1903.
22. *Commercial Telegraphers' Journal*, 1914.
23. *Commercial Telegraphers' Journal*, August 1907.
24. *Telecommunications: A Program for Progress*, report by the President's Communications Policy Board, Washington, D.C., March 1951, p. 99.

Conclusion

1. *Journal of the Telegraph*, February 1, 1869.
2. *Harpers New Monthly Magazine*, March 1873, p. 358.
3. These self-winding clocks were connected by wire to master clocks in Western Union's time service departments. These master clocks are the ones regulated from the National Observatory signals, and are then used to automatically correct the clocks connected to its circuit.
4. Lubrano, Annteresa, *The Telegraph: How Technological Innovation Caused Social Change*, p. 120.
5. E.B. White, *Quo Vadimus? Or, The Case for the Bicycle* (New York: Harper and Bros., 1939), p. 13.
6. W.J. Johnson, *Telegraph Tales and Telegraph History*, pp. 53, 144, 157.
7. *Telegraph and Telephone Age*, June 1, 1932.

Bibliography

Periodicals, Papers and Government Records

"An Act to Provide for the Incorporation and Regulation of Telegraph Companies." New York *Laws*, April 12, 1848.

Andrews, J. Cutler. "The Southern Telegraph Company, 1861–1865: A Chapter in the History of Wartime Communication," *The Journal of Southern History*, Vol. 30, No. 3 (August 1964).

Annual Report of the President of the Western Union Telegraph Company, 1869.

Annual Report of the President of the Western Union Telegraph Company, 1873.

Annual Report of the President of the Western Union Telegraph Company, 1880.

Annual Report of the President of the Western Union Telegraph Company, 1887.

"The Atlantic Telegraph," *The Commercial and Financial Telegraph Chronicle*, 1 (July 8, 1865) 34.

Banquet Speech, April 15, 1864, in "Europe and America: Report on the Proceedings at an Inauguration Banquet Given by Mr. Cyrus Field." London: William Brown, 1864. Box 142 (Cyrus Field Papers), Collection #205 (Western Union Telegraph Company), 1993, Addendum, National Museum of American History.

Bektas, Yakup. "The Sultan's Messenger: Cultural Constructions of Ottoman Telegraphy, 1847–1880," *Technology and Culture*, Vol. 41, No. 4 (October 2000).

Carpenter, Frank G. "Henry Clay on Nationalizing the Telegraph," *The North American Review*, Vol. 154, No. 424 (March 1892).

Congressional Globe, 39th Congress, 1st Session (1866), pp. 979–980.

Constitution, Brotherhood of Telegraphers, Sec. 1, Article 15.

Cooper, Peter. "Reminiscences."

Craig, Daniel. Printed circular to managers of telegraph companies (May 1851).

Deposition of W.H. Seward, July 7, 1870, RG 123 (Records of the U.S. Court of Claims), General Jurisdiction, Case File 6151, Box 306, National Archives and Records Administration (NARA).

"Diplomacy as a Profession," *National Review* (London) 35 (March 1900).

du Boff, Richard B. "Business Demand and the Development of the Telegraph in the United States, 1844–1860," *Business History Review*, 54 (Winter 1980).

du Boff, Richard B. "The Telegraph in Nineteenth Century America: Technology and Monopoly," *Comparative Studies in Society and History*, Vol. 26, No. 4 (October 1984).

Dunne, Peter Finley. "Mr. Dooley on Diplomatic Indiscretions," New York *Times*, March 22, 1914.

"The Electric Telegraph," *Daily Chronicle*, November 15, 1867.

Federal Communications Commission, "Corporate History of the Western Union Telegraph Company," 1915; 73rd

Bibliography

Congress, 2nd Session, House Report No. 1273, Part 3, No. 4 (1935) 3967–74, referred to as the Splawn Report.

Field, James Alexander. "The Magnetic Telegraph, Price and Quantity Data and the New Management of Control," *The Journal of Economic History*, Vol. 52, No. 2 (June 1992), p. 413.

Fifth Annual Report of the Pennsylvania Railroad Company to the Stockholders, 1851.

Final Report and Testimony, Commission on Industrial Relations, April 1914, Vol. 10.

Holcombe, A.M. "Public Ownership of Telegraphs and Telephones," *The Quarterly Journal of Economics*, Vol. 28, No. 3 (May 1914).

Hubbard, Gardner G. "The Proposed Changes in the Telegraphic System," *North American Review*, 117 (July 1873).

Huebner, S.S. "The Functions of Produce Exchanges," *Annals of the American Academy of Political and Social Science*, 38 (September 1911), pp. 319–353.

"Influence of the Telegraph on Commerce," *Hunt's Merchant Magazine*, 59 (August 1868).

"Intellectual Effects of Electricity," *London Spectator*, Nov. 9, 1889.

Jepsen, Thomas C. "Women Telegraph Operators on the Western Front," *Journal of the West*, 35 (2): 1996, p. 75.

Jones, Fred M. "The Middleman in the Domestic Trade of the United States, 1800–1860," *Illinois Studies in Social Sciences*, Vol. 21, No. 3 (Urbana, IL, 1937), p. 15.

Kennedy, P.M. "Imperial Cable Communications and Strategy, 1870–1914," *English Historical Review*, 86 (1971).

"The Magnetic Telegraph West," *Public Ledger and Daily Transcript* (Philadelphia), August 24, 1846.

Nickles, David, Paull. "Telegraph Diplomats: The United States Relations with France in 1848 and 1870," *Technology and Culture*, Vol. 40, No. 1 (January 1999).

Phelps, Edward John. "International Relations: Address before the Phi Beta Kappa Society of Harvard University," Burlington, VT: Free Press Association, June 29, 1889.

Postmaster General's Report, 1890.

Pound, Roscoe. "The Military Telegraph in the Civil War," *Proceedings of the Massachusetts Historical Society*, Third Series, Vol. 66 (October 1939–May 1941).

Radcliffe, Viscount Strafford de, May 31, 1861, in British Parliamentary Papers, Report from the Select Committee on Diplomatic Service, Reports from Committees, 1861, 6: p. 168.

Reed, James D. "The Western Press and the Telegraph," *National Telegraph Review* 1 (October 1853).

Report of Hearings, U.S. Industrial Commission, 1914.

Report of the Committee of the Senate upon the Relations between Labor and Capital, and Testimony Taken by the Committee (Washington, D.C.: Government Printing Office, 1885) 1: 101–21, 1: 185–97.

"Report of the Directors of the New York and Erie Railroad to the Stockholders," *American Railroad Journal*, Vol. 27, January 14, 1854.

Report of the Proceedings of a Meeting of the Stockholders of the Atlantic and Ohio Telegraph Company, July 19, 1849.

Rothstein, Morton. "Antebellum Wheat and Cotton Exports: A Contrast in Marketing Organization and Economic Development," *Agricultural History*, 40 (April 1966).

Samuel F.B. Morse Papers. Library of Congress, M33, fr. 365.

Schwantes, Benjamin Sidney Michael. "Fallible Guardian: The Social Construction of Railroad Telegraphy in 19th Century America." Dissertation, University of Delaware, Fall 2008.

Senate Committee on Post Office and Post Roads, Report on Postal Tele-

Bibliography

graph, 48th Congress, 1st Session, 1884, S. Report 577, 1 and 5.

Shanks, William F. S. "How We Get Our News," *Harpers New Monthly Magazine* (May 1967).

Shaw, Donald L. "News Bias and the Telegraph: A Study of Historical Change," *Journalism Quarterly*, Spring 1967.

"Sketch of the Life of J.H. Wade from 1811 to About 1867," Jepha H. Wade Papers, Series 1, Container 1, Folder 2, *Western Reserve Historical Society*, Cleveland, OH.

Smith, Andy. "Sun Sets on 150 Years of the Foreign Office Cable," *Observer*, August 16, 1998.

"The Southern Telegraph," *New Orleans Commercial Times*, July 13, 1847.

"Statements of O.H. Palmer, Alonzo Cornell, and Geo. Walker, Vice Presidents of Western Union," *New York Times*, Jan. 5, 1870.

"Telecommunications: A Program for Progress, A Report by the President's Communication Policy Board," Washington, D.C., March 1951.

"The Telegraph," *Republican* (St. Louis), November 23, 1847.

Testimony of President William Orton, 41st Congress, 2nd Session, House Report No. 114 (July 5, 1870), p. 118.

United States Bureau of Corporations, "Report of the Commissioner of Corporations on the Beef Industry," March 3, 1905 (Washington, D.C., 1905), p. 207.

United States Congress, House of Representatives, 41st Congress, 2nd Session (1869–1870) House Report No. 114 (serial volume 1438), "The Postal Telegraph in the United States." Washington, D.C.: Government Printing Office, 1870.

United States Congress, Senate, Investigation of Western Union and Postal Telegraph Companies, 60th Congress, 2nd Session, S. Doc. 725. Washington, D.C., 1909, pp. 21–22.

The War of the Rebellion: A Compilation of the Official Records of the Union and Confederate Armies, Series 1, 17, Part 2, pp. 385–386.

Western Union Report to the Stockholders, 1873.

Wisconsin Press Association, Proceedings, 1872.

Books

Abbot, Edith. *Women in Industry: A Study in American Economic History.* New York: Source Book Press, 1970.

Adams, Charles Francis. *Memoirs of John Quincy Adams*, 12 vols. Philadelphia: J.B. Lippincott, 1874.

Aiken, John. *Labor and Wages, at Home and Abroad.* Lowell, MA: D. Bixby, 1849.

Albion, Robert G. *The Rise of the New York Port.* New York: Charles Scribner's Sons, 1939.

Allen, Zachariah. *Science of Mechanics.* Providence: Hutchens and Cory, 1829.

Ammon, Harry. *The Great Mission.* New York: Norton, 1973.

Appleton, Nathan. *Introduction of the Power Loom and Origin of Lowell.* Lowell, MA: printed by B.H. Penhallow, 1858.

Appleton, Nathan. *Labor: Its Relations in Europe and in the United States Compared.* Boston: Eastburn's Press, 1844.

Arnold, R.A. *History of the Cotton Famine.* London: Saunders, Otley, 1865.

Audubon, John James. *John James Audubon: Writings and Drawings.* New York: Library of America, 1999.

Bagnall, William R. *The Textile Industry of the United States, Including Sketches and Notes of Cotton, Wollen, Silk, and Linen Manufacturers in the Colonial Period.* Boston: W.B. Clarke, 1893.

Baldwin, Leland, D. *The Keelboat Age on Western Waters.* Pittsburgh: University of Pittsburgh Press, 1941.

Bartlett, Elisha. *A Vindication of the Character and Condition of the Females Employed in the Lowell Mills.* Lowell, MA: L. Huntress, 1841.

Bibliography

Batchelder, Samuel. *Introduction and Early Progress of Cotton Manufacture in the United States*. Boston: Little, Brown, 1863.

Bateman, Fred, and Thomas Weiss. *A Deplorable Society: The Failure of Industrialization in the Slave Economy*. Chapel Hill: University of North Carolina Press, 1981.

Bemis, Samuel Flagg. *Jay's Treaty: A Study in Commerce and Diplomacy*. New York: Macmillan, 1923.

Betts, Edwin Morris, ed. *Thomas Jefferson's Farm Book: With Commentary and Relevant Extracts from Other Writings*. Philadelphia: American Philosophical Society, 1944.

Blumenthal, Henry. *A Reappraisal of Franco-American Relations, 1830–1871*. Chapel Hill: University of North Carolina Press, 1959.

Bolles, Albert. *Industrial History of the United States*. Norwich, CT: H. Bell, 1879.

Bowen, Catherine Drinker. *Miracle at Philadelphia: The Story of the Constitutional Convention, May to September 1787*. Boston: Little, Brown, 1966.

Boyd, Julian P. *The Papers of Thomas Jefferson*. Princeton, NJ: Princeton University Press, 1950.

Bruchey, Stuart. *Cotton and the American Economy: 1790–1860*. New York: Harcourt, Brace and World, 1967.

Cairnes, J. E. *The Slave Power: Its Character, Career, and Probable Design: Being an Attempt to Explain the Real Issues Involved in the American Contest*. New York: Kelles, 1968.

Caldwell, Stephen A. *A Banking History of Louisiana*. Baton Rouge: Louisiana State University Press, 1935.

Callender, Guy S., ed. *Selections from the Economic History of the United States, 1765–1860*. Boston: Ginn, 1909.

Carey, H.C. *Essay on the Rate of Wages*. New York: Kelley, 1965.

Carlson, Avery Luvere. *A Monetary and Banking History of Texas*. Fort Worth: Fort Worth National Bank, 1930.

Catterall, Ralph C. *Second Bank of the United States*. Chicago: University of Chicago Press, 1960.

Chestnut, Mary Boykin. *A Diary from Dixie*. Ben Ames Williams, ed. Boston: Houghton, Mifflin, 1949.

Clark, Victor S. *History of Manufactures in the United States*, 2nd ed., 1. New York: McGraw-Hill, 1929.

Clayton, A.S. *Compilation of the Laws of Georgia*. Augusta: Adams and Duyckinck, 1812.

Cohen, I. Bernard. *Benjamin Franklin's Science*. Cambridge: Harvard University Press, 1990.

Cohn, David L. *The Life and Times of King Cotton*. New York: Oxford University Press, 1956.

Cooper, Thomas, and D.J. McCord. *South Carolina Statutes at Large*. Columbia: A.S. Johnson, 1836–1841.

Crockett, David. *The Life of David Crockett*. New York: Charles Scribner's Sons, 1923.

Crowley, John E. *The Privileges of Independence: Neomerchantilism and the American Revolution*. Baltimore: Johns Hopkins Press, 1993.

Dattel, Gene. *Cotton and Race in the Making of America*. Chicago: Ivan R. Dee, 2009.

Davis, Charles S. *The Cotton Kingdom in Alabama*. Philadelphia: Porcupine Press, 1974.

Davis, Jefferson. *The Rise and Fall of the Confederate Government*, 2 vols. New York: T. Yoseloff, 1958.

DeBow, J.D.B. *The Industrial Resources of the Southern and Western States*, 3 vols., 2. New Orleans: pub. at the offices of De Bow's Review, 1852–1853.

De Pauw, Linda G., et al. *Documentary History of the First Federal Congress of the United States of America*, Vol. 1: *Senate Legislative Journal*. Baltimore: Johns Hopkins University Press, 1972.

Dobson, John M. *History of America En-*

Bibliography

terprise. Upper Saddle River: Prentice Hall, 1988.

Donnell, E.J. *History of Cotton*. Wilmington, DE: Scholarly Resources, 1973.

Drake, Samuel Adams. *History of Middlesex County, Massachusetts*, 2 vols. Boston: Estes and Laureat, 1880.

Dunbar, Rowland, ed. *Jefferson Davis, Constitutionalist: His Letters, Papers, and Speeches*, 10 vols. Jackson, MS: Torgerson Press, 1923.

Dutton, H.I. *The Patent System and Inventive Activity During the Industrial Revolution, 1750–1852*. Manchester, U.K.: Manchester University Press, 1984.

Ellison, Thomas. *The Cotton Trade of Great Britain*. New York: A.M. Kelley, 1968.

Ellison, Thomas. *The History of Great Britain*. London: E. Wilson, 1886.

Eltis, David. *Economic Growth and the Ending of the Transatlantic Slave Trade*. New York: Oxford University Press, 1987.

Eno, Arthur L. *Cotton Was King: A History of Lowell, Massachusetts*. New Hampshire: New Hampshire Publishing Co., 1976.

Evans, Estwick A. *Pedestrious Tour of Four Thousand Miles*. Cleveland: Arthur H. Clark, 1904.

Farnie, D.A. *The English Cotton Industry and the World Market, 1815–1896*. Oxford: Clarendon Press, 1979.

Ferguson, Eugene S. *On the Origin and Development of American Mechanical Know-how*, Cambridge, MA: MIT Press, 1992.

Fitton, R.S., and A.P. Wadsworth. *The Strutts and the Arkwrights, 1758–1830: A Study of the Early Factory System*. Manchester, U.K.: Manchester University Press, 1973.

Fogel, Robert William. *Without Consent or Contract: The Rise and Fall of American Slavery*. New York: Norton, 1989.

Fogel, Robert William, and Stanley L. Engerman. *Time on the Cross: The Economics of American Negro Slavery*. New York: W.W. Norton, 1974.

Fox-Genovese, Elizabeth, and Eugene D. Genovese. *Fruits of Merchant Capital*. New York: Oxford University Press, 1983.

Gibb, George S. *Saco-Lowell Shops: Textile Machinery in New England*, Cambridge: Harvard University Press, 1950.

Goodrich, Carter, et al. *Canals and American Economic Development*. New York: Columbia University Press, 1961.

Gray, Lewis Cecil. *History of Agriculture in the Southern United States to 1860*, Vol. 1. Glouster: Peter Smith, 1958.

Green, Constance McLoughlin. *Holyoke, Massachusetts*. New Haven: Yale University Press, 1939.

Green, George D. *Finance and Economic Development in the Old South: Louisiana Banking, 1804–1861*. Stamford: Stamford University Press, 1972.

Hall, James. *The West*. Cincinnati: H. W. Derby, 1848.

Hamilton, Alexander. *Report on Manufacturers*. London: J. Debrett, 1793.

Hamilton, Alexander, and Henry Cabot Lodge. *The Works of Alexander Hamilton*. New York: G.P. Putnam's Sons, 1904.

Hamilton, Joseph Gregoire de Roulhac. *The Papers of Thomas Ruffin*. Raleigh: Edwards and Broughton, 1918–1920.

Helper, Hinton B. *Compendium of the Impending Crisis of the South*. New York: A.B. Burdick, 1860.

Howe, Daniel. *The Political Culture of American Whigs*. Chicago: University of Chicago, 1979.

Hunter, Louis C. *Steamboats on the Western Rivers*. Cambridge: Harvard University Press, 1949.

Iken, Arthur. *Texas*. Waco: Texian Press, 1964.

Jefferson, Thomas. *Notes on Virginia*. Richmond: J.W. Randolph, 1853.

Jones, Fred Mitchell. *Middlemen in the*

Bibliography

Domestic Trade of the United States. Urbana: University of Illinois, 1937.

Josephson, Hannah. *The Golden Threads: New England Mill Girls and Magnates*. New York: Duell, Sloan, and Pearce, 1949.

Kasson, John F. *Civilizing the Machine: Technology and Republican Values in America, 1776–1900*. New York: Grossman, 1976.

Kerber, Linda K. *Women of the Republic: Intellect and Ideology in Revolutionary America*. Chapel Hill: University of North Carolina Press, 1980.

Knowlton, Evelyn H. *Pepperell's Progress: History of a Cotton Textile Company, 1844–1945*. Cambridge: Harvard University Press, 1948.

Kohn, August. *The Cotton Mills of South Carolina*. Spartanburg: Reprint Co., 1975.

Labree, Leonard W. *The Papers of Benjamin Franklin*, Vol. 6. New Haven: Yale University Press, 1959.

Lipscomb, Andrew A., and Albert Ellery Bergh, eds. *The Writings of Thomas Jefferson*. Washington, D.C.: Thomas Jefferson Memorial Association of the United States, 1903–1904.

Loring, F.W., and C.W. Atkinson. *Cotton Culture and the South Considered with Reference to Emigration*. Boston: A. Williams, 1869.

Loughbridge, R.H. *Cotton Production in Georgia*. Washington, D.C.: Government Printing Office, 1884.

Lowrie, Walter, and M. St. Clair Clarke. *American State Papers: Documents, Legislative and Executive*. Washington, D.C: Gales and Seaton, 1832.

Luther, Seth. *An Address to the Working Men of New England on the State of Education and on the Condition of Producing Classes in Europe and America*. 2nd ed., Boston: the author, 1832.

MacLeod, Christine. *Inventing the Industrial Revolution: The English Patent System, 1660–1800*. Cambridge: Cambridge University Press, 2007.

Marbury, H., and W. H. Crawford. *Compilation of the Laws of Georgia*. Savannah: Seymour, Woolhopter, and Stebbins, 1802.

Marrs, Aaron W. *Railroads in the Old South, Pursuing Progress in a Slave Society*. Baltimore: Johns Hopkins University Press, 2009.

Martin, Edger, W. *The Standard of Living in 1860: American Consumption Levels on the Eve of the Civil War*. Chicago: University of Chicago Press, 1942.

McEwan, Barbara. *Thomas Jefferson: Farmer*. Jefferson, NC: McFarland, 1991.

McKissick, James R. *Notes on the Early History of Cotton and Cotton Manufacturers in South Carolina*. Spartanburg: Band and White, 1927.

McMaster, J.B. *History of the People of the United States*. New York: Farrar, Straus, 1964.

McPherson, James M. *Battle Cry of Freedom: The Civil War Era*. New York: Oxford University Press, 1988.

Meier, August, and Elliot Rudwick. *From Plantation to Ghetto*. New York: Hill and Wang.

Miller, John C. *Alexander Hamilton, Portrait in Paradox*. New York: Harper, 1959.

Miller, Randall Martin. *The Cotton Mill Movement in Antebellum Alabama*. New York: Arno Press, 1971.

Mims, Shadrach. *History of Autauga County*. Prattville, AL: A.R.B.C., 1976.

Mims, Shadrach. *History of Prattville*. Prattville, AL: A.R.B.C., 1976.

Mirsky, Jeannette, and Allan Nevins. *The World of Eli Whitney*. New York: Macmillan, 1952.

Mitchell, Broadus. *The Industrial Revolution in the South*. New York: New American Library, 1930.

Moore, Frank, ed. *The Rebellion Record: A Diary of American Events*. New York: G.P. Putnam, 1861–1863.

Necker, Jacques. *A Treatise on the Ad-*

Bibliography

ministration of Finance. London: Logographic Press, 1787.

Neel, Joanne Loewe. *Phineas Bond: A Study in Anglo-American Relations*. Philadelphia: University of Pennsylvania Press, 1968.

Nemmo, Joseph. *Report on the Internal Commerce of the United States*. Washington, D.C.: Government Printing Office, 1885.

Nicolay, John G., and John Hay, eds. *Abraham Lincoln: Complete Works*, 2 vols. New York: F. D. Tandy, 1905.

Nuermberger, Ruth K. *The Free Produce Movement: A Quaker Protest Against Slavery*. Durham: Duke University Press, 1942.

Olmsted, Frederick Law. *A Journey to the Back Country*. New York: Schocken Books, 1970.

Onuf, Peter, and Nicholas Onuf. *Federal Union, Modern World: The Law of Nations in an Age of Revolution*. Madison: Madison House, 1993.

Ousley, Frank L. *King Cotton Diplomacy*. Chicago: University of Chicago Press, 1959.

Parker, William N. *Europe, America, and the Wider World*. Cambridge, MA: Cambridge University Press, 1984.

Phelps, Ulrich B. *American Negro Slavery*. New York: D. Appleton, 1918.

Pitkin, T.A. *A Statistical View of the Commerce of the United States*. New York: Augustus M. Kelly, 1967.

Ramsey, David. *History of South Carolina*. Columbia, SC: W.J. Duffie, 1858.

Reid, James. D. *The Telegraph in America: Its Founders, Promoters and Noted Men*. New York: Derby, 1879.

Richard, Henry. *Memoirs of Joseph Sturge*. London: S.W. Partridge, 1864.

Robinson, Harriet. *Loom and Spindle or Life Among the Early Mill Girls*. Kailua, HI: Press Pacifica, 1976.

Russell, Robert. *North America: Its Agriculture and Climate*. Edinburgh: Black, 1857.

Scherer, James A.B. *Cotton as a World Power: A Study in the Economic Interpretation of History*. New York: Frederick A. Stokes, 1916.

Schoolcraft, Mary Howard. *The Black Gauntlet*. Freeport, NY: Books for Librarians Press, 1971.

Setser, Vernon. *The Commercial Reciprocity Policy of the United States, 1774–1829*, New York: Da Capo Press, 1937.

Shore, Laurence. *Southern Capitalists: The Ideological Leadership of an Elite, 1832–1855*. Chapel Hill: University of North Carolina Press, 1986.

Silver, Arthur W. *Manchester Men and Indian Cotton*. Manchester, U.K.: Manchester University Press, 1966.

Sobel, Robert. *The Money Manias: Tales of Entrepreneurs and Investors During the Eras of Great Speculation in America, 1770–1970*. New York: Weybright and Talley, 1973.

Stanwood, Edward. *American Tariff Controversies in the Nineteenth Century*, 2 vols. New York: Houghton Mifflin, 1903.

Stevens, William. *A Journey of the Proceedings in Georgia, Beginning October 20, 1737*, 3 vols.

Stover, John F. *American Railroads*. Chicago: University of Chicago Press, 1961.

Sumner, Helen. *History of Women in Industry in the United States*. New York: Arno Press, 1974.

Sutcliff, Robert. *Travels in Some Parts of North America in the Years 1804, 1805, and 1806*. Philadelphia: B&T Kite, 1812.

Syndor, Charles S. *Slavery in Mississippi*. Glouster, MA: P. Smith, 1965.

Syrett, Harold G. *Papers of Alexander Hamilton*. New York: Columbia University Press, 1976.

Tarrant, Susan F.H. *Hon. Daniel Pratt: A Biography with Eulogies on His Life and Character*. Richmond: Whittet and Shepperson, 1904.

Taussig, Frank W. *Some Aspects of the*

Bibliography

Tariff Question. Cambridge: Harvard University Press, 1915.

Taussig, Frank W. *Tariff History of the United States.* London: G.P. Putnam's Sons, 1931.

Taylor, John. *Arator*, 6th ed. Georgetown: J.M. Carter, 1814.

Thompson, William T. *Major Jones' Sketches of Travels.* Charlottesville: University of Virginia, 2000.

Thorp, Willard L. *Business Annals.* New York: National Bureau of Economic Research, 1926.

Tindall, George B. *The Emergence of the New South, 1913-1946.* Baton Rouge: Louisiana State University Press, 1967.

Tocqueville, Alexis de. *Democracy in America.* New York: A.A. Knopf, 1945.

Walton, Perry. *The Story of Textiles.* Boston: J.S. Laurence, 1912.

Ware, Caroline. *Early New England Cotton Manufacture.* New York: Russell and Russell, 1966.

Watkins, J.L. *King Cotton.* New York: Negro University Press, 1969.

Watts, John. *The Facts of the Cotton Famine.* London: Cass, 1968.

Webster, Noah. *A Grammatical Institute of the English Language*, Part 1. Albany: Charles R. and George Webster, 1796.

Wentworth, Joseph, and George Wallis. *The Industry of the United States in Machinery Manufacturers and Useful and Ornamental Arts.* London: G. Routledge, 1854.

White, George. *Statistics of Georgia.* Savannah: W. Thorne Williams, 1849.

White, Laura A. *Robert Barnwell Rhett: Father of Secession.* Gloucester, MA: P. Smith, 1931.

"Williams to Anson Jones," in *Anson Jones, Republic of Texas.* Chicago: Rio Grande Press, 1966.

Wilson, Clyde N. "Calhoun's Economic Platform," in *Slavery, Secession, and Southern History*, Robert Louis Paquette and Louis A. Ferleger, eds. Charlottesville: University of Virginia Press, 2000.

Winks, Robin W. *Canada and the United States: The Civil War Years.* Baltimore: Johns Hopkins Press, 1960.

Wright, Richardson. *Hawkers and Walkers in Early America.* Philadelphia: J.B. Lippincott, 1927.

Yafa, Stephen. *Big Cotton.* New York: Viking, 2005.

York, Neil Longley. *Mechanical Metamorphosis: Technological Change in Revolutionary America.* Westport, CT: Greenwood Press, 1985.

Zevin, Robert B. "The Growth of Cotton Production After 1815," in *The Reinterpretation of America's Past*, Robert Fogel and Stanley Engermann, eds. New York: Harper and Row, 1974.

Index

Adams, Charles Francis 47
Agamemmon 94, 96
American Rapid Telegraph Company 178–179
"American System of train dispatching" 116
Anglo-American Telegraph Company 107
Associated Press 9, 128
Atlantic Cable 95, 97–99, 108

Block System (railroad safety) 114
Blondheim, Menahem 134
Brassy, Thomas 102
Brett's Magnetic Telegraph Company 102
Brotherhood of Telegraphers 176
Burleson Gen. Albert S. 160

Candler, Alfred 119
carrier pigeon 9, 136
Caton, Judge J.D. 185
Clay, Henry 149–150
Colfax, Schuyler 159
Colt, Samuel 13
Cooper, Peter 90, 92
Craig, Daniel 136–144

Daniell, John Frederic 2
Daniell cell 2
diplomacy: impact of telegraph on 34; lessening of power of foreign ministers 42–46

Eckert, Thomas 75
Edison, Thomas 153, 159, 178
Elliembic 1

Field, Cyrus 9, 22, 89
Fleming, Sanford 9
Franco-Prussian War 35
Franklin, Benjamin 1

Galbraith, John Kenneth 34
Gisborne, Frederick Newton 89
Glass, Elliot and Company 101
Gobright, Augustus 141
Gold Reporting Telegraph 124
Golden, H.H. 153
Gompers, Samuel 183–184
Gooch, Daniel 102
Gould, Jay 157–158
Grant, Ulysses S. 59–61, 63–65, 82–83
Gray, Harrison 1
Great Eastern 93, 103
Green, Norvin 180

Hallock, Gerald 142
Hamlin, Cyrus 39
Harbor News Association 128
Haupt, Gen. Herman 113
hazards of being a telegraph operator 71
Headrick, Daniel 32–33
Henry, Joseph 1
Hill, Senator Nathaniel P. 160
Hochfelder, David 158–159
Hooker, Gen. Joseph 80–81
House Royal 19
Hubbard, Gardner 11
Hughes, David 23
Hull Cordell 47

International Telegraphic Union 129

Jackson, Gen. Stonewall 77–79
Johnson, W.J. 167, 187

Kendall, Amos 1, 17, 162
Knights of Labor 176, 180

Lincoln, Abraham: Atlantic Telegraph 101; obtaining military information via telegraph 80; political use of telegraph

Index

86; speech as president-elect 141; using telegraph at beginning of war 75–76
Lindley, Lester 170
Lindqvist, Svante 49
Little, Jacob 135

Mackay, John W. 158
Magnetic Telegraph Company 91
Mallet, Frank 35
Maury, Lt. Matthew Fontaine 90–91
McCallum, Daniel C. 116–117
McClellan, Gen. George 51, 101
Meade, Gen. George 81
Medill, Joseph 142
Merrimac 77
Morse, Samuel: Atlantic Cable 94, 96; claim to invention of telegraph 6; electro-magnetism 5; invests in telegraph 150; original idea 2; proves to Congress telegraph works 14; railroad safety 115–116; tests telegraph 13

National Telegraphic Union 172
New York City (importance to the telegraph) 130
news: foreign 132; by wire vs. by mail 133
North Atlantic Telegraph Company 99–100

"Orders in Council" 29–30
O'Reilly, Henry 17, 19–21, 136–137
Orton, William 26, 129, 152, 159–160, 162–164, 166–167
Orwell, George 41
Ottoman Empire 39–40

Parsons, Frank 182
pashas 39
Polk, Pres. James 32, 42
Post Roads Act 152–155
Postal Telegraph Company 128, 158
"postalization" 161
Press Association War of 1866/1867 129
Pupin, Michael 159

railroads 110–114, 118
Railway Gazette 116
Reid, James D. 131, 133
Reuter's News Agency 101
Roche, William 69–70
Rosecrans, Gen. William 81–82
Rush, Richard 30–32

Seward, Sen. William Henry 36, 93
Seymour, Charles 147
Shrimpton, B.F. 171
Sibley, Hiram 151
Smith, Francis O.J. 134–138
Society for the Diffusion of Political Knowledge 24
Southern Telegraph Association 70–71
Southern Telegraph Company 71–72
Spaulding, Bishop M.J. 22–23
Stager, Anson 53
Stanton, Edwin 50
strikes 173–181, 183–184
Sullivan, William 176

telegraph: Confederate telegraph situation at beginning of Civil War 66–70; confidentiality 123; cost of use 35–36; cotton trade and 122–123; department stores and 122; development by Samuel Morse 5–6; diplomacy 8; effect on commodity exchanges 122; effect on diplomatic authority 38; espionage 8; first completed line 6; garbled communications 37; government control 6; intercity communication 124–125; late delivery 46–47; late message deliveries 46–47; meat packers and 122; newspapers and 8; operator problems during Civil War 56–62, 65–66; praised by Civil War Union generals 64–65; private lines 124; railroads and 7; security 36–37; single line problems 130–132; telegraph/railroad sample contract 118; Union use during Civil War 62; warfare and 8
The Telegraph in America 127
(The) Telegrapher 172
Telegraphers Protective League 173–175
time 186–187
Trent Affair 100–101

unions (telegraphers) 168–169
United States Military Telegraph 168–169

Vail, Alfred 149
Van Creveld, Martin 130
Vanderbilt, William 157
Van Duzer, J.C. 53, 55
The Vatican 49

212